Target Tracking and Navigation for Intelligent Autonomous Unmanned Systems Application

Target Tracking and Navigation for Intelligent Autonomous Unmanned Systems Application

Guest Editors

Chunhui Zhao
Shuai Hao

Basel • Beijing • Wuhan • Barcelona • Belgrade • Novi Sad • Cluj • Manchester

Guest Editors

Chunhui Zhao
Northwestern Polytechnical University
Xi'an
China

Shuai Hao
Xi'an University of Science and Technology
Xi'an
China

Editorial Office
MDPI AG
Grosspeteranlage 5
4052 Basel, Switzerland

This is a reprint of the Special Issue, published open access by the journal *Sensors* (ISSN 1424-8220), freely accessible at: https://www.mdpi.com/journal/sensors/special_issues/98406ISZWU.

For citation purposes, cite each article independently as indicated on the article page online and as indicated below:

Lastname, A.A.; Lastname, B.B. Article Title. *Journal Name* **Year**, *Volume Number*, Page Range.

ISBN 978-3-7258-3113-5 (Hbk)
ISBN 978-3-7258-3114-2 (PDF)
https://doi.org/10.3390/books978-3-7258-3114-2

© 2025 by the authors. Articles in this book are Open Access and distributed under the Creative Commons Attribution (CC BY) license. The book as a whole is distributed by MDPI under the terms and conditions of the Creative Commons Attribution-NonCommercial-NoDerivs (CC BY-NC-ND) license (https://creativecommons.org/licenses/by-nc-nd/4.0/).

Contents

About the Editors vii

Shuai Hao, Kang Ren, Jiahao Li and Xu Ma
Transmission Line Defect Target-Detection Method Based on GR-YOLOv8
Reprinted from: *Sensors* **2024**, *24*, 6838, https://doi.org/10.3390/s24216838 1

Xu Ma, Tianqi Li, Jun Deng, Tong Li, Jiahao Li, Chi Chang, et al.
Infrared and Visible Image Fusion Algorithm Based on Double-Domain Transform Filter and Contrast Transform Feature Extraction
Reprinted from: *Sensors* **2024**, *24*, 3949, https://doi.org/10.3390/s24123949 15

Siya Sun, Sirui Mao, Xusheng Xue, Chuanwei Wang, Hongwei Ma, Yifeng Guo, et al.
Research on Obstacle-Avoidance Trajectory Planning for Drill and Anchor Materials Handling by a Mechanical Arm on a Coal Mine Drilling and Anchoring Robot
Reprinted from: *Sensors* **2024**, *24*, 6866, https://doi.org/10.3390/s24216866 37

Jingyi Du, Ruibo Zhang, Rui Gao, Lei Nan and Yifan Bao
RSDNet: A New Multiscale Rail Surface Defect Detection Model
Reprinted from: *Sensors* **2024**, *24*, 3579, https://doi.org/10.3390/s24113579 56

Yongbin Su, Chenying Lin and Tundong Liu
Real-Time Trajectory Smoothing and Obstacle Avoidance: A Method Based on Virtual Force Guidance
Reprinted from: *Sensors* **2024**, *24*, 3935, https://doi.org/10.3390/s24123935 72

Jianqun Yao, Jinming Li, Yuxuan Li, Mingzhu Zhang, Chen Zuo, Shi Dong and Zhe Dai
A Vision–Language Model-Based Traffic Sign Detection Method for High-Resolution Drone Images: A Case Study in Guyuan, China
Reprinted from: *Sensors* **2024**, *24*, 5800, https://doi.org/10.3390/s24175800 86

Zhan-Wu Ma and Wan-Sheng Cheng
Visual-Inertial RGB-D SLAM with Encoder Integration of ORB Triangulation and Depth Measurement Uncertainties
Reprinted from: *Sensors* **2024**, *24*, 5964, https://doi.org/10.3390/s24185964 112

Shi Lei, He Yi and Jeffrey S. Sarmiento
Synchronous End-to-End Vehicle Pedestrian Detection Algorithm Based on Improved YOLOv8 in Complex Scenarios
Reprinted from: *Sensors* **2024**, *24*, 6116, https://doi.org/10.3390/s24186116 131

Jingzhao Zhang, Yuxin Cui, Zhenguo Yan, Yuxin Huang, Chenyu Zhang, Jinlong Zhang, et al.
Time Series Prediction of Gas Emission in Coal Mining Face Based on Optimized Variational Mode Decomposition and SSA-LSTM
Reprinted from: *Sensors* **2024**, *24*, 6454, https://doi.org/10.3390/s24196454 148

Yu Du, Jianjun Hao, Zijing Chen and Yijun Guo
Indeterministic Data Collection in UAV-Assisted Wide and Sparse Wireless Sensor Network
Reprinted from: *Sensors* **2024**, *24*, 6496, https://doi.org/10.3390/s24196496 164

About the Editors

Chunhui Zhao

Chunhui Zhao is a Professor and Doctoral Supervisor at Northwestern Polytechnical University. Their research interests include UAV autonomous navigation control, collaborative positioning, intelligent games, and fusion decision making. Having conducted research on major national topics, such as the National Natural Science Foundation of China, and a special project on civil aircraft, Chunhui Zhao has published five academic monographs and over sixty papers in the fields of visual perception, autonomous navigation, and information fusion. A member of the National Aircraft Data Definition and Management sub-technical Committee, they have also won provincial- and ministerial-level prizes in first place, as well as other science and technology awards.

Shuai Hao

Shuai Hao, Ph.D., is an Associate Professor and a Masters supervisor. He serves as Vice Chairman of the 8th Electrical Engineering Education Professional Committee of the China Electrotechnical Society, and is a senior member of the Chinese Association of Automation. Dr. Hao has led over 20 projects, including those funded by the National Natural Science Foundation of China, the China Postdoctoral Science Foundation, and the Shaanxi Provincial Science and Technology Plan Project. He has published more than 50 papers in domestic and international journals and conferences, with over 40 being indexed in SCI and EI. Additionally, he has obtained over 10 invention patents and 5 utility model patents. His research interests include target detection, fault diagnosis in electrical equipment, and artificial intelligence.

Article

Transmission Line Defect Target-Detection Method Based on GR-YOLOv8

Shuai Hao [1], Kang Ren [1,2], Jiahao Li [1] and Xu Ma [1,*]

[1] College of Electrical and Control Engineering, Xi'an University of Science and Technology, Xi'an 710054, China; haoxust@163.com (S.H.); kk2580256@163.com (K.R.); 18406060425@stu.xust.edu.cn (J.L.)
[2] Huaneng Longdong Energy Co., Ltd., Qingyang 744500, China
* Correspondence: maxu@xust.edu.cn

Abstract: In view of the low levels of speed and precision associated with fault detection in transmission lines using traditional algorithms due to resource constraints, a transmission line fault target-detection method for YOLOv8 (You Only Look Once version 8) based on the Rep (Representational Pyramid) Visual Transformer and incorporating an ultra-lightweight module is proposed. First, the YOLOv8 detection network was built. In order to address the needs of feature redundancy and high levels of network computation, the Rep Visual Transformer module was introduced in the Neck part to integrate the pixel information associated with the entire image through its multi-head self-attention and enable the model to learn more global image features, thereby improving the computational speed of the model; then, a lightweight GSConv (Grouped and Separated Convolution, a combination of grouped convolution and separated convolution) convolution module was added to the Backbone and Neck to share computing resources among channels and reduce computing time and memory consumption, by which the computational cost and detection performance of the detection network were balanced, while the model remained lightweight and maintained its high precision. Secondly, the loss function Wise-IoU (Intelligent IOU) was introduced as the Bounding-Box Regression (BBR) loss function to optimize the predicted bounding boxes in these grid cells and shift them closer to the real target location, which reduced the harmful gradients caused by low-quality examples and further improved the detection precision of the algorithm. Finally, the algorithm was verified using a data set of 3500 images compiled by a power-supply inspection department over the past four years. The experimental results show that, compared with the seven classic and improved algorithms, the recall rate and average precision of the proposed algorithm were improved by 0.058 and 0.053, respectively, compared with the original YOLOv8 detection network; the floating-point operations per second decreased by 2.3; and the picture detection speed was increased to 114.9 FPS.

Keywords: transmission line defect detection; vision transformer; YOLOv8; light weight; loss function

1. Introduction

In recent years, owing to rapid economic development, the total length of high-voltage transmission lines worldwide has exceeded four million kilometers. In order to ensure the safety and quality of transmission, power line inspections have received increasingly more attention throughout the world [1]. Due to their long-term exposure to harsh external environments such as high winds and heavy rains, there has been high frequency of faults in transmission lines [2–4]. Therefore, regular transmission line inspections are very important to ensuring their safety and stability. UAV inspection has been widely used in power line inspections because of its flexibility, high speed, and safety. However, the massive amount of inspection data cannot be timely processed due to limited equipment resources, etc. Consequently, many detection algorithms are not readily usable, lacking

the required speed and precision, which causes a large workload [5]. Therefore, given the limited nature of the equipment resources, achieving accurate and efficient defect defection from the image data set generated in UAV inspection is of great practical significance.

At present, algorithms based on deep learning theory have made significant progress in the field of target detection. This approach is mainly based on a data-driven method for target detection. It has the advantages of high detection accuracy, strong robustness, and multi-target detection capabilities. This is a research hotspot in computer vision. The relevant deep learning algorithms can be roughly divided into two categories: One is the two-stage algorithm, which is represented by R-CNN [6] (Region-based Convolutional Neural Networks), Fast R-CNN [7], and FPN [8] (Feature Pyramid Network). Wang et al. [4] used the R-CNN algorithm to locate the spacers, shock absorbers, and isobaric rings on the UAV inspection images, and realized the real-time and accurate identification of these transmission line components. Liang et al. [9] proposed a transmission line defect detection method based on deep learning, using Faster R-CNN to construct a transfer learning detection model. This method has good robustness for defect detection under different illumination conditions. However, the two-stage algorithm is often computationally complex, as characterized by its detailed candidate region generation and classification process detection, resulting in its slow speed and difficulty in meeting the detection requirements involved with massive data.

The other category comprises a one-stage algorithm, including the SSD [10] (Single Shot MultiBox Detector) and YOLO [11] (You Only Look Once) series, as well as RetinaNet [12], EfficientDet [13], and other algorithms. These algorithms have obvious advantages in speed and are more suitable for large-scale detection tasks that require high levels of efficiency. By extracting high-quality features from aerial images and using multi-level perception, the advantages characteristic of global and local information can achieve a favorable balance and effectively improve the detection accuracy [14]. The method designed by Hao et al. [15] improves the feature extraction ability of the network and performs effective feature fusion. However, the applicability of the detection algorithm to the inspection of transmission lines is not considered in the above research, and graphics cards and processors associated with the transmission line inspection equipment are seldom timely updated. As a result, the precision and speed of many algorithms are not achieved because of these inadequate software and hardware configurations, resulting in the failure of the timely processing of massive amount of defect-related image data generated in inspections.

In view of the above problems, a YOLOv8 multi-fault target-detection method for transmission lines is proposed based on Visual Transformers and ultra-lightweight modules, one which is simply referred to as GR-YOLOv8. The main innovations and contributions of this paper are as follows:

(1) As to the obvious high requirements inherent in the network computing and feature map redundancy of the YOLOv8 algorithm, a Rep Visual Transformer module is introduced in the Neck part, and the multi-head self-attention module is used to enhance the model's ability to learn image features and reduce the feature dimensions to be processed and the computational burden, thereby improving the computational speed of the model.

(2) Give the generally limited equipment resources and the low applicability of the detection algorithm, the lightweight GSConv convolution module was introduced in the Backbone part; in this module, an optimization strategy for the combination of grouped convolution and separated convolution is adopted to reduce the amount of computation and reduce memory loss. Computing resources can also be dynamically allocated based on actual hardware performance to balance computing costs and detection performance, which not only achieves higher detection precision, but also reduces power and energy consumption. Especially in terms of small-size feature maps, GSConv convolution creates a lower computational burden than does standard convolution, which improves the precision as to small targets to certain extent.

(3) The loss function CIoU was modified to Wise-IoU. After considering the influence of the position of the bounding box, the tolerance of the model for position errors was adjusted to optimize the final test result. In addition to the reduction in the competitiveness of high-quality anchor frames, the harmful gradients generated by low-quality examples were also reduced. Therefore, the bounding-box regression was more accurate, which further improves the detection precision of the model.

2. Principle of YOLOv8 Detection Algorithm

The YOLOv8 network architecture includes three key parts (as shown in Figure 1): the Backbone network for feature extraction, the Backbone network for feature fusion, and the detection head for the final identification and detection.

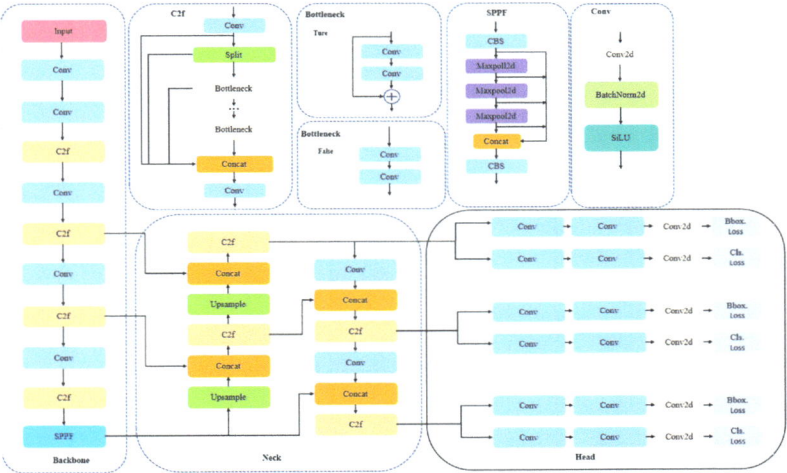

Figure 1. YOLOv8 network structure.

The Backbone network is a local cross-stage network designed to reduce the computational load and optimize gradient propagation. In order to capture spatial information more efficiently, it integrates the spatial pyramid pooling module [16]. Compared to YOLOv5's PAN-FPN, the Neck network eliminates the convolution of the upsampling stage. Instead, downsampling is performed prior to the upsampling. At the same time, the C3 model is replaced with a lighter C2f module [17], which enhances the model's adaptability as to targets of different sizes and shapes. The detection head utilizes the currently popular decoupling structure, which not only reduces the number of parameters and computational complexity, but also improves the generalization and stability of the model.

The loss of the bounding box of the YOLOv8 algorithm is calculated by the Conv2d layer. At the same time, the traditional anchor-based prediction method is abandoned, and replaced with an anchor-free strategy aiming to directly predict the center coordinates and aspect ratio of the target, thereby minimizing anchor boxes and improving the detection speed and precision.

3. GR Lightweight Model

Classic lightweight models include MobileNets [18,19], ShuffleNets [20–22], and Ghost-Net [23–25]. As one example, the core feature of MobileNets is the use of a special Depth-wise Separate Convolution (DSC), in which the traditional convolution is divided into two independent functions of depthwise convolution and pointwise convolution. The depthwise convolution is very efficient in processing high-resolution images, while pointwise convolution involves the application of 1×1 convolution on each channel separately, a process which actually consists of the global-average pooling of all channels before the expansion is expanded. Therefore, the effect of pointwise convolution is similar to that of the feature fusion of channels. However, 1 × 1 intensive convolution takes up more computing resources, and often leads to lower levels of detection precision. In addition, many lightweight models use only DSC in their basic architecture design, from the beginning to the end, which causes the direct amplification of the defects of DSC in the Backbone network. Therefore, precision has been significantly reduced, regardless of whether it is used for image classification or detection.

In order to make the output of DSC close to scattering (SC), a new method was introduced with the use of a mixed convolution of SC, DSC, and shuffle; this was called Grouped Spatial Convolution 9 GSConv [26]. This convolution preserves these connections as much as possible with lower time complexity, especially in mobile devices or circumstances where resources are limited. Therefore, GSConv reduced the amount of computing and memory loss, and improved the detection of small targets. The time complexity formulas of SC, DSC, and GSConv are as follows:

$$Time_{SC} \sim 0(W \cdot H \cdot K_1 \cdot K_2 \cdot C_1 \cdot C_2) \tag{1}$$

$$Time_{DSC} \sim 0(W \cdot H \cdot K_1 \cdot K_2 \cdot 1 \cdot C_2) \tag{2}$$

$$Time_{GSConv} \sim 0\left(W \cdot H \cdot K_1 \cdot K_2 \cdot \frac{C_2}{2} \cdot (C_1 + 1)\right) \tag{3}$$

where Time is time complexity; W is the width of the output feature map; H is the height of the output feature map; K1·K2 is the size of the convolution kernel; C1 is the number of channels in each convolution kernel, as well as the number of channels in the input feature map; and C2 is the number of channels in the output feature map. The structure of the GSConv model is shown in Figure 2.

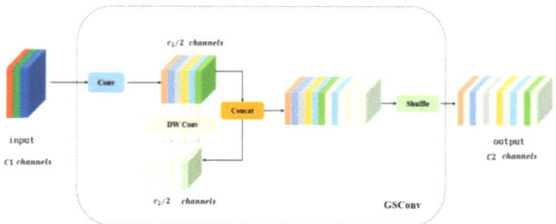

Figure 2. GSConv network structure.

Shuffle is used for the penetration of the SC information into the various parts of the DSC information. This method allows SC information to be evenly exchanged with local characteristic information on different channels and completely mixed into the DSC output without cumbersome steps, which not only optimizes the model performance and reduces computational complexity and memory consumption but also maintains a certain degree of model expression, especially in mobile devices or circumstances where resources are limited.

In addition, RepViT [27] utilizes an improved Transformer architecture, one which combines traditional convolutional networks with Visual Transformer (ViT). In particular, it introduces a multi-head self-attention module to enhance the model's ability to learn image features. The multi-head self-attention mechanism allows the model to process information associated with different scales and locations in parallel. The overall architecture of the RepViT model is obtained through 14 optimization steps, such as the alignment of the training formula, delay of metric, the alignment of the training strategy, the optimization of the block design, and the separation of the token mixer. The structure is shown in Figure 3.

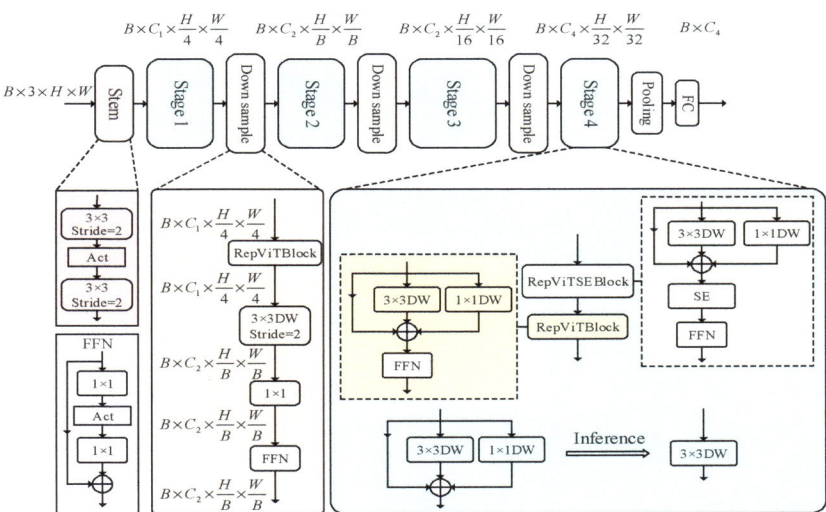

Figure 3. The overall architecture of the RepViT model.

In this architecture, H and W are the width and height of the input image; Ci is the size of the i-th channel; and B is the size of the batch.

In the end, the Rep Visual Transformer module was introduced in the Neck part, and the lightweight GSConv convolution module was added to the Backbone and Neck. The lightweight design allows the model to reduce computing time and memory consumption without excessively sacrificing precision and is suitable for real-time applications or object detection on a resource-limited device.

4. Wise-IoU Loss Function

The bounding-box loss function plays a vital role in improving the model's target-detection speed. Since current research focuses on improving the matching of precise bounding boxes, the over-processing of many common poor-quality samples in the actual training data may adversely affect the detection effect.

The original YOLOv8 uses GIOU [28] as the loss function of the bounding box. Although it solves the failure to predict the distance between the intersection over union (IOU) prediction box and the real box when they do not intersect, GIOU cannot measure their relative positional relationship when the prediction box and the real box are in an inclusive relationship. As a result, the detection effect of the model is affected.

Therefore, the Wise-IoU [29] loss function is introduced herein. In this function, with the adoption of the two core strategies of Feature Pyramid Networks (FPN) and finer regional proposal generation technology, low-level feature maps were used for detection of small objects while high-level feature maps were used for detection of large objects, which allows the capture of more potential target information. At the same time, the loss function evaluates the quality of the anchor frame based on its outlier. By identification of the small offset of the real bounding box of the predicted object relative to the preset candidate area, the prediction box can cover the target object more accurately, and reduce the negative gradient effect of low-quality samples, without overemphasizing the high-quality anchor frame. Therefore, Wise-IoU has the significant ability to process medium-quality anchor frames, which improves the overall performance of the detector. The calculation formula for Wise-IoU is:

$$L_{WIoU} = R_{WIoU} \times L_{IoU} L_{WIoUv1} = R_{WIoU} \times L_{IoU} \qquad (4)$$

$$R_{WIoU} = exp\frac{\left(b_{c_x}^{gt} - b_{c_x}\right)^2 + \left(b_{c_y}^{gt} - b_{c_y}\right)^2}{c_w^2 + c_h^2} \quad (5)$$

$$L_{IoU} = 1 - IoU \quad (6)$$

The parameters of Equations (4)–(6) are shown in Figure 4. $\left(b_{c_x}^{gt} - b_{c_x}\right)^2$ is the Euclidean distance between the center points of the real box and the prediction box; h and w are the height and width; and c_h and c_w are the height and width of the smallest closed box composed of the prediction box and the real box.

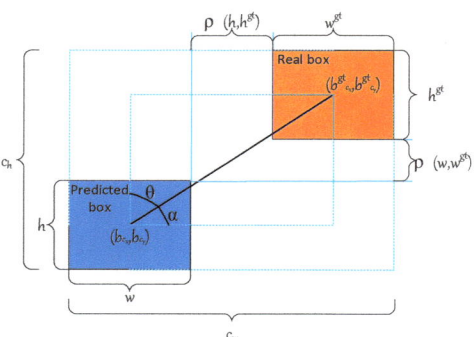

Figure 4. Schematic diagram of the parameters of the loss function.

5. Experiments and Analysis of Results

Innovative improvements were made to the target-detection algorithm of the YOLOv8 model, including the application of Rep Visual Transformer and lightweight GSConv module, and the optimization of the loss function into Wise-IoU, etc. The structural diagram of the proposed GR-YOLOv8 network is shown in Figure 5.

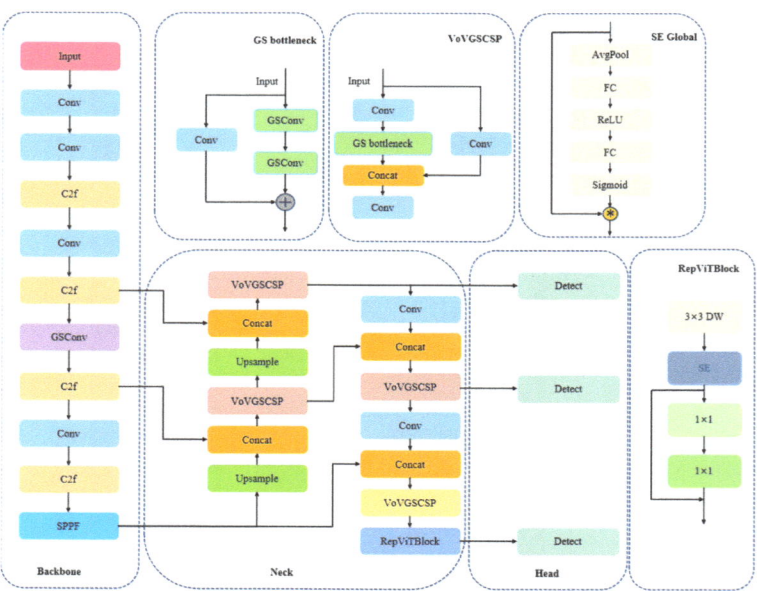

Figure 5. GR-YOLOv8 network structure diagram.

In order to simulate an environment with limited on-site equipment resources, the most common on-site hardware and software configurations were adopted: Windows 10 operating system, NVIDIA GeForce GTX1050 graphics card (with 4 GB video memory), Intel (R) Core (TM) i5-8300H, Pytorch 3.8, and PyCharm 2021.2.4.

5.1. Collection and Pre-Processing of Experimental Data

The image data associated with 330 kV high-voltage transmission lines and acquired by inspections using drones conducted by the on-site department of a power supply administration agency over the past 4 years were used. The 12 common problems with transmission lines included wire damage, insulator shedding, bird nest formation and spacer rod disconnection. Some defect targets are shown in Figure 6.

(a) Wire damage

(b) The insulator falls off

(c) Nest formation

(d) The spacer rod is disconnected

Figure 6. Common defects in transmission lines.

A data set of 3500 inspection images from a power supply administration agency was selected, and their defects were marked in detail with LabelImg. The classifications and numbers of the defects are shown in Table 1.

Table 1. Defect name and quantity.

Fault Type	Custom Name	Quantity
Pin having come off	XT	342
Wire damage	DXS	316
Nut having come off	LST	347
Discharge gap too large	FDG	310
Insulator sheds	JYT	212
Bird's nest	NC	235
Gap design problems	FDSJ	262
Anti-bird spur damage	NS	231
Gap shorting	FDD	310
Gaskets missing	DP	323
Spacer bar disconnected	JGB	364
Sundry issues	ZW	248

5.2. Model Training

The various anomalies in the data set were randomly divided into training sets and test sets at proportions of 70% and 30%, respectively. During the model training, the size of all image samples was adjusted to 640 × 640 to ensure their consistency and reduce the computational burden caused by the original size difference. The stochastic gradient descent optimization algorithm was used to update the model parameters, and the total number of iterations was set to 100 times with a batch size of 16, which is suitable for scenarios in which a quick response to data streams is expected with limited resources. The initial learning rate was set to 0.01. The learning rate determines the update speed of the model parameters. If it is set too high, it may lead to unstable training, and if it is set too low, it may lead to slow convergence. The momentum used was 0.937. This optimization strategy not only considers the direction of the current gradient in each iteration, but also retains a part of the memory of the previous gradient direction, which helps to smooth the update and reduce oscillations, which is especially significant in dealing with considerable local minima.

5.3. Model Evaluative Indicators

The evaluative indicators of the model were precision (P), recall (R), floating-points (Flops), average precision (AP), mean average precision (mAP), and frame rate per second (FPS). The calculation formulas associated with some of the evaluative indicators are:

$$P = \frac{N_{TP}}{N_{TP} + N_{FP}} \times 100\% \tag{7}$$

$$R = \frac{N_{TP}}{N_{TP} + N_{FN}} \times 100\% \tag{8}$$

$$AP = \int_0^1 P(R)dR \tag{9}$$

$$mAP = \frac{1}{n}\sum_{i=1}^{n} AP_i \tag{10}$$

$$FPS = \frac{1}{t} \tag{11}$$

P is the proportion of positive examples among those predicted to be positive by the model; R is to the proportion of actual positive examples correctly recognized by the model; N_{TP} is the positive sample accurately predicted by the algorithm; N_{FP} is the positive sample misjudged by the algorithm; N_{FN} is the positive sample not predicted by the algorithm; Flops is the number of floating-points required during the operation of the model, which is used to measure the computational complexity and hardware resource consumption of the model; the average precision (AP) is determined by calculating the area under the P–R curve, and the higher the value, the better the performance of the model in identifying a specific target; mAP is the average of the APs of all categories; and n is the total number of categories. For the i-th category of targets, the larger the mAP, the higher the precision of the algorithm. FPS is used to measure the detection speed of the algorithm, which refers to the average time it takes to detect a frame of image. The larger the FPS, the higher the detection speed and the better the real-time performance.

5.4. Analysis of Results

The real-time change curve of the loss function during the model training is shown in Figure 7.

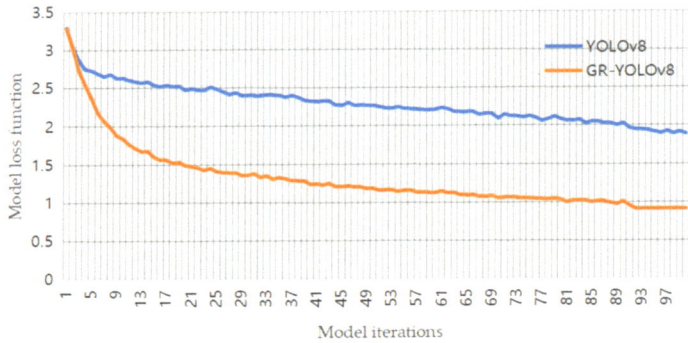

Figure 7. Loss function curve.

Figure 7 shows the comparison of the training processes of the GR-YOLOv8 and YOLOv8 algorithms. In the initial stage, the loss values of the two are close, but in subsequent iterations, the loss value of the proposed algorithm drops faster and more steadily, and is always at a low level. In the end, the loss value of GR-YOLOv8 was approximately 0.9, while the value for YOLOv8 was approximately 1.9. From the experimental results in Figure 6, it can be seen that GR-YOLOv8 algorithm achieves better training results.

The mAP@0.5 curves of the GR-YOLOv8 and YOLOv8 algorithms are shown in Figure 8.

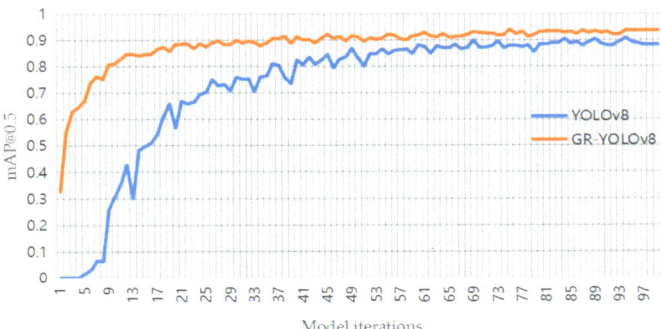

Figure 8. mAP@0.5 contrast curve.

It can be seen from Figure 8 that after 20 cycles of training iterations, the mAP@0.5 of GR-YOLOv8 algorithm rose to about 0.9. Comparatively, after 70 cycles of training iterations, the mAP@0.5 of YOLOv8 rose to only about 0.9, and this level was not sustained for long. In the end, the mAP@0.5 of GR-YOLOv8 stabilized at about 0.935, and that of YOLOv8 stabilized at about 0.882. Therefore, compared with the YOLOv8 algorithm, the proposed algorithm is improved to a certain extent in terms of the mean of average precision.

From the test results in Table 2, it can be seen that the average precision of the proposed GR-YOLOv8 algorithm reached 0.935 and the recall rate reached 0.922.

Table 2. Detection results for the GR-YOLOv8 model on various targets.

Defect Type	Precision	Recall
XT	0.955	1
DXS	0.974	0.939
LST	0.976	0.970
FDG	0.995	1
JYT	0.953	0.943
NC	0.969	0.964
FDSJ	0.995	1
NS	0.643	0.500
FDD	0.911	0.957
DP	0.948	0.914
JGB	0.956	0.939
ZW	0.946	0.939
all	0.935	0.922

In order to verify the impacts of the use of Gsconv, RepViT, and Wise-IoU on the performance of the model, ablation experiments were conducted based on the original yolov8 algorithm, as shown in Table 3.

Table 3. Ablation experiments.

Gsconv	RepViT	Wise-IoU	mAP	FPS	Parameter Quantity
-	-	-	0.882	68.7	3,740,727
√	-	-	0.894	90.3	3,185,239
√	√	-	0.907	114.9	2,850,881
√	√	√	0.935	114.9	2,850,881

In Table 3, "-" means "not adopted", and "√" means "adopted". In the first line, the mAP of the original YOLOv8 algorithm was 0.872. After improvements in three aspects, the number of parameters decreased by 889,846, and the FPS was greatly improved. The mAP reached 0.935, which was 0.053 higher than that of the original YOLOv8 algorithm. It can be seen that the lightweight design of GR-YOLOv8 algorithm model achieves a very obvious effect, and the detection precision has also been improved accordingly. In order to further objectively evaluate the advantages of the GR-YOLOv8 algorithm, some classic and improved algorithms were compared through experiments. All algorithms were trained using the above data sets and configurations. The results are shown in Table 4.

Table 4. Performance comparisons of different algorithms.

Model	FLOPS	Recall	mAP	FPS
YOLOv8	9.2	0.864	0.882	68.7
EMA	8.1	0.852	0.889	74.1
simEMA	9.1	0.843	0.849	68.9
Swin Transformer	9.8	0.908	0.917	64.1
DSConv	21.9	0.085	0.077	52.1
Detection head	15.4	0.912	0.925	28.3
Gsconv	8.4	0.876	0.894	87.2
GR-YOLOv8	6.9	0.922	0.935	114.9

As shown in the comparative experimental results in Table 4, the floating-points, recall rate, mean of average precision, and detection rate of YOLOv8 were 9.2, 0.864, 0.882, and 68.7 FPS, respectively. Then, the EMA [30,31] (Efficient Multi-Scale Attention in Cross-Space) and simEMA (Simplified Exponential Moving Average Attention) attention mechanisms were tried, for which the core C2 module was replaced with C2f-EMA and C2f-simEMA. Although this change slightly improves the image detection speed and floating-point computing speed, after the introduction of the EMA module, the recall rate was reduced by 0.012 compared to YOLOv8. After the application of the simEMA module, the recall rate was reduced by 0.021, and the average precision was also significantly reduced by 0.033.

As to the Swin Transformer algorithm, its architecture design is similar to that of the traditional convolution hierarchy, in which the resolution of each layer is halved but the number of channels is doubled. The principle is to divide the image into windows, that is, the division of input sequence into windows of fixed size or dynamic size. Then, a standard self-attention mechanism is used to calculate local features inside each window, followed by interaction and spicing of these windows. The results show that compared with YOLOv8, although the algorithm had improved the recall rate by 0.044 and the mean of average precision by 0.035, the number of floating-point operations increased significantly, and the image detection speed also decreased by 4.6 FPS.

Dynamic Snake Convolution (DSConv) adopts the precise segmentation of a tubular structure to ensure the precision and efficiency of downstream tasks. Hence, it was tried in the application of defect target-detection for transmission lines. However, because most defects are at the towers and poles of transmission lines, the experimental results were too unsatisfactory in terms of floating-point calculation, recall rate, mean of average precision, and image detection speed. After the addition of the detection head module, the ability of detection as to small targets was greatly improved, and the recall rate and average precision were increased by 0.048 and 0.043, respectively, although the image detection speed decreased significantly.

The detection performance of the GR-YOLOv8 algorithm was significantly improved. The recall rate and the mean of average precision increased by 0.058 and 0.053, respectively. The floating-point operations per second decreased by 2.3, and the image detection speed increased to 114.9 FPS. With its lightweight design and ease of on-site application in an environment with resource constraints, the detection precision is improved while the computational complexity and memory consumption are reduced, which verifies the feasibility of the improvement plan.

In order to verify the advantages of the proposed algorithm as to detection, four groups of scenes were selected for the testing, including those with multiple categories, partial occlusion, small scale, and complex backgrounds. In Figure 9a, these are arranged from left to right. The defects in the images were marked with rectangular boxes. The comparison results for YOLOv8, EMA, Swin Transformer, DSConv, detection head, Gsconv, and GR-YOLOv8 detection algorithms in the experiments are shown in Figure 9.

As shown in Figure 9, the defects of the transmission lines were clearly identified, and the detection confidence of YOLOv8 was between 0.7 to 1.0, while values for other algorithms fluctuated between 0.3 and 1.0. It is particularly obvious that the Depthwise Separate Convolution (DSConv) failed to identify the defects in some cases. In contrast, the GR-YOLOv8 algorithm exhibits a high level of confidence, which ensures that defects are precisely identified in most cases. In addition, the algorithm in this paper can detect a variety of defects of transmission lines in a single image, which shows satisfactory adaptability, stability, and robustness. The analysis of the four sets of test results is conducted in the following paragraphs (Group 1 is on the left).

There are many types of defects in the Group 1 experiments, and they are largely different in shape and size. The Dynamic Snake Convolution (DSConv) performed poorly in dealing with multiple types of defects and failed to achieve the necessary precision. Other algorithms have a high degree of responsiveness in the detection of multiple types of defects, and without obvious missed detections. Among them, the detection results of the YOLOv8, Swin Transformer, detection head, and Grouped Spatial Convolution (Gsconv) algorithms were consistent. The confidence for the detection results of Exponential Moving Average (EMA) was low, and the GR-YOLOv8 algorithm can more precisely identify and locate multiple types of defects, achieving better detection results.

The Group 2 experiments involve the detection of defects in complex environments. The experimental results show that YOLOv8, EMA, Swin Transformer, and DSConv all missed detections when dealing with complex backgrounds, which indicates that these algorithms are not responsive to defects against complex backgrounds. The GR-YOLOv8 detection algorithm had a high response rate in this experiment, and the confidence of the detection results was also high, 0.7 and 0.8, respectively, which indicates better performance in complex environments.

The Group 3 experiments involve the detection of insulator defects, which is critically important in the defect inspection of a power system. The experimental results show that the detection results for the Depthwise Separate Convolution (DSConv) were low, with a confidence level of only 0.4, and its ability in feature extraction and the handling of specific types of defects is limited to certain extent, a limitation which requires further optimization and improvement as to the detection of insulator defects. The confidence of the other six algorithms in the detection of insulator defects was 1, which shows that they achieve higher precision and reliability and better detection performance.

Group 4 experiments involve the detection of small defects. The DSConv algorithm exhibits obvious false-negative cases, which demonstrates its obvious disadvantages in this respect. The confidence of the detection results for the EMA and Swin Transformer algorithms was only 0.3, and that of the YOLOv8 algorithm was 0.9. Comparatively, the confidence of the detection results of GR-YOLOv8 algorithm was 1.0. Therefore, it can precisely identify and locate small defects, showing better detection results.

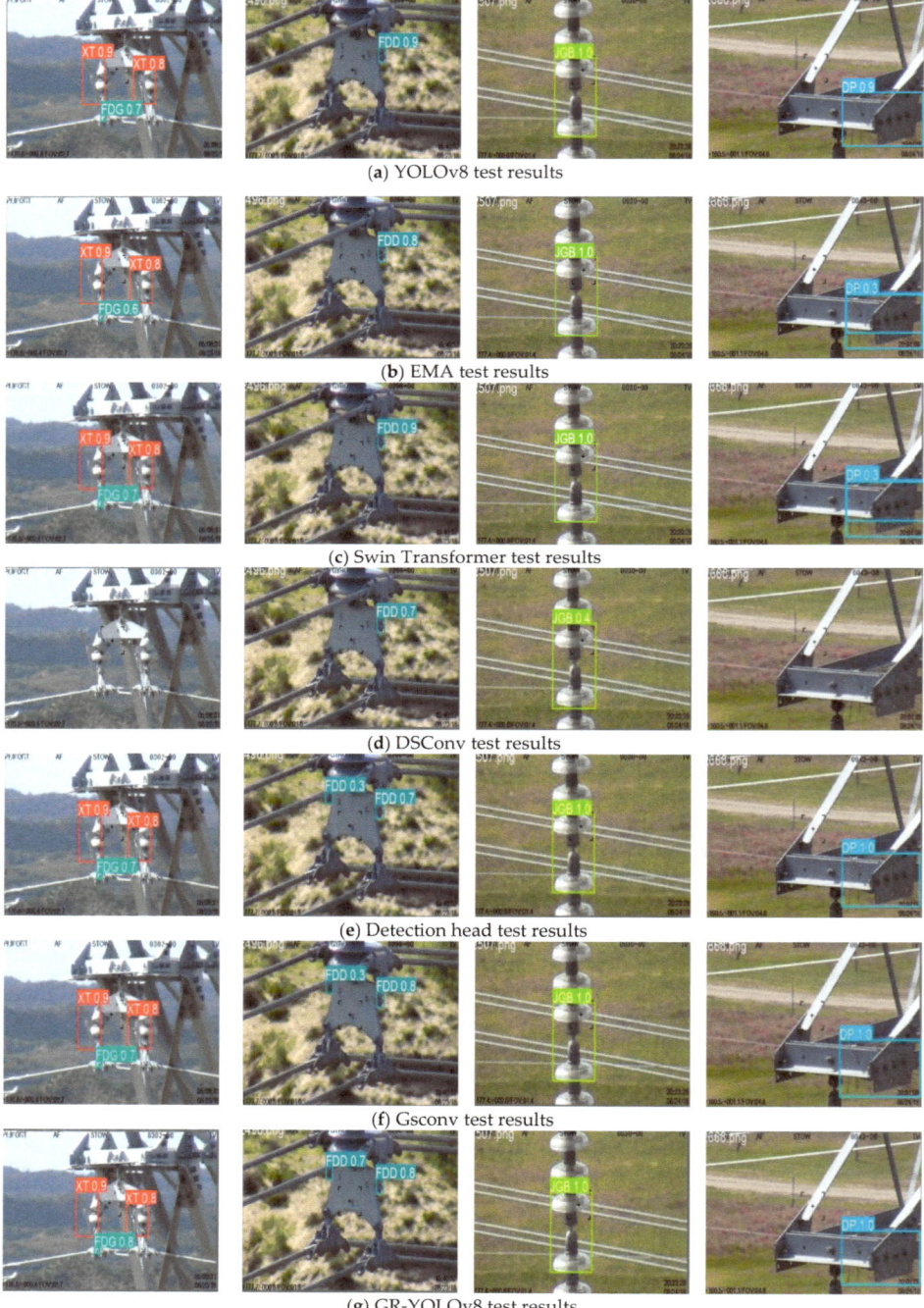

(a) YOLOv8 test results

(b) EMA test results

(c) Swin Transformer test results

(d) DSConv test results

(e) Detection head test results

(f) Gsconv test results

(g) GR-YOLOv8 test results

Figure 9. Comparison of detection results.

6. Conclusions

In view of the present low levels of speed and precision in the detection of defects in transmission lines with traditional algorithms and given situations in which resources are limited, a YOLOv8 target fault-detection method for transmission lines is proposed, based on Visual Transformers and ultra-lightweight modules. The Rep Visual Transformer module was introduced in the Neck part to integrate the pixel information for the entire image through its multi-head self-attention. This enabled the model to learn more global image features, which improved the computational speed of the model and thus solved for feature redundancy and high network computing demands; the lightweight GSConv convolution module was added to the Backbone and Neck to share computing resources among channels and reduce computing time and memory consumption, and the computational cost and detection performance of the detection network were thereby balanced to improve the detection precision and speed, especially in the detection of small targets. At the same time, the loss function Wise-IoU was introduced to reduce the harmful effects caused by low-quality examples, and further improved the detection precision of the algorithm. Compared with seven classic or improved high-speed algorithms such as EMA, Swin Transformer, and Detection head, GR-YOLOv8 can reduce the number of parameters and calculations, as well as the memory usage, in circumstances involving limited software and hardware resources, which reduces computing time and energy consumption while maintaining a sufficient degree of feature extraction. This balances computing costs and detection performance in commonly used equipment, greatly improving the detection speed while improving the precision in defect detection to certain extent.

Author Contributions: Conceptualization, S.H. and K.R.; methodology, K.R. and J.L.; software, K.R.; validation, S.H., K.R. and J.L.; formal analysis, X.M.; investigation, K.R.; resources, S.H.; data curation, S.H. and X.M.; writing—original draft preparation, K.R.; writing—review and editing, K.R. and J.L.; visualization, K.R.; supervision, S.H., K.R., J.L. and X.M.; project administration, S.H.; funding acquisition, S.H. and X.M. All authors have read and agreed to the published version of the manuscript.

Funding: This work was funded by the China Postdoctoral Science Foundation under Grant 2020M683522, and the Natural Science Basic Research Program of Shanxi under Grant 2024JC-YBMS-490.

Institutional Review Board Statement: Not applicable.

Informed Consent Statement: Not applicable.

Data Availability Statement: Restrictions apply to the availability of these data. Data were obtained from power grid bureau and are available from the authors with the permission of power grid bureau.

Conflicts of Interest: Author Kang Ren was employed by the company Huaneng Longdong Energy Co., Ltd. The remaining authors declare that the research was conducted in the absence of any commercial or financial relationships that could be construed as a potential conflict of interest.

References

1. Yang, L.; Fan, J.; Liu, Y.; Li, E.; Peng, J.; Liang, Z. A Review on State-of-the-Art Power Line Inspection Techniques. *IEEE Trans. Instrum. Meas.* **2020**, *69*, 9350–9365. [CrossRef]
2. Zhao, Z.; Wang, R.; Li, Y.; Zhai, Y.; Zhao, W.; Zhang, K. A new multilabel recognition framework for transmission lines bolt defects based on the combination of semantic knowledge and structural knowledge. *IEEE Trans. Instrum. Meas.* **2022**, *71*, 1–11. [CrossRef]
3. Zhao, Z.; Guo, G.; Zhang, L.; Li, Y. A new anti-vibration hammer rust detection algorithm based on improved YOLOv7. *Energy Rep.* **2023**, *9*, 345–351. [CrossRef]
4. Zhao, W.; Kang, Y.; Chen, H.; Zhao, Z.; Zhai, Y.; Yang, P. A target detection algorithm for remote sensing images based on a combination of feature fusion and improved anchor. *IEEE Trans. Instrum. Meas.* **2022**, *71*, 1–8. [CrossRef]
5. Liu, X.; Miao, X.; Jiang, H.; Chen, J. Box-point detector: A diagnosis method for insulator faults in power lines using aerial images and convolutional neural networks. *IEEE Trans. Power Deliv.* **2021**, *36*, 3765–3773. [CrossRef]
6. Girshick, R.; Donahue, J.; Darrell, T. Region-based convolutional networks for accurate object detection and segmentation. *IEEE Trans. Pattern Anal. Mach. Intell.* **2016**, *38*, 142–158. [CrossRef]

7. Ren, S.; He, K.; Girshick, R. Faster R-CNN: Towards truth-time object detection with region proposal networks. *IEEE Trans. Pattern Anal. Mach. Intell.* **2015**, *39*, 1137–1149. [CrossRef]
8. Lin, T.-Y.; Dollár, P.; Girshick, R.; He, K.; Hariharan, B.; Belongie, S. Feature pyramid networks for object detection. In Proceedings of the IEEE Conference on Computer Vision and Pattern Recognition, Honolulu, HI, USA, 21–26 July 2017; pp. 2117–2125.
9. Liang, H.; Zuo, C.; Wei, W. Detection and evaluation method of transmission line defects based on deep learning. *IEEE Access* **2020**, *8*, 38448–38458. [CrossRef]
10. Liu, W.; Anguelov, D.; Erhan, D.; Szegedy, C.; Reed, S.; Fu, C.Y.; Berg, A.C. SSD: Single shot multibox detector. In *Computer Vision—ECCV 2016*; Springer: Cham, Switzerland, 2016; pp. 21–37.
11. Redmon, J.; Divvala, S.; Girshick, R. You only look once: Unified, ruth-time object detection. In Proceedings of the 2016 IEEE Conference on Computer Vision and Pattern Recognition (CVPR), Las Vegas, NV, USA, 27–30 June 2016; pp. 779–788.
12. Lin, T.; Goyal, P.; Girshick, R.; He, K.; Dollar, P. Focal loss for dense object detection. In Proceedings of the IEEE International Conference on Computer Vision (ICCV), Venice, Italy, 22–29 October 2017; pp. 2999–3007.
13. Tan, M.; Pang, R.; Le, Q.V. EfficientDet: Scalable and efficient object detection. In Proceedings of the IEEE/CVF Conference on Computer Vision and Pattern Recognition, Seattle, WA, USA, 13–19 June 2020; pp. 10781–10790.
14. Jiang, H.; Qiu, X.J.; Chen, J.; Liu, X.; Miao, X.; Zhuang, S. Insulator Fault Detection in Aerial Images Based on Ensemble Learning With Multi-Level Perception. *IEEE Access* **2019**, *7*, 61797–61810. [CrossRef]
15. Hao, K.; Chen, G.; Zhao, L.; Li, Z.; Liu, Y.; Wang, C. An insulator defect detection model in aerial images based on multiscale feature pyramid network. *IEEE Trans. Instrum. Meas.* **2022**, *71*, 3522412. [CrossRef]
16. Song, X.; Fang, X.; Meng, X.; Fang, X.; Lv, M.; Zhuo, Y. Real-time semantic segmentation network with an enhanced backbone based on Atrous spatial pyramid pooling module. *Eng. Appl. Artif. Intell.* **2024**, *133*, 107988. [CrossRef]
17. Jin, Y.; Tian, X.; Zhang, Z.; Liu, P.; Tang, X. C2F: An effective coarse-to-fine network for video summarizatio. *Image Vis. Comput.* **2024**, *144*, 104962. [CrossRef]
18. Bi, C.; Wang, J.; Duan, Y.; Fu, B.; Kang, J.-R.; Shi, Y. MobileNet Based Apple Leaf Diseases Identification. *Mob. Netw. Appl.* **2022**, *27*, 172–180. [CrossRef]
19. Chang, H.-C.; Yu, L.-W.; Liu, B.-Y.; Chang, P.-C. Classification of the implant-ridge relationship utilizing the MobileNet architecture. *J. Dent. Sci.* **2024**, *19*, 411–418. [CrossRef]
20. Chen, Z.; Yang, J.; Chen, L.; Jiao, H. Garbage classification system based on improved ShuffleNet v2. *Resour. Conserv. Recycl.* **2022**, *178*, 106090. [CrossRef]
21. Yu, Y.-N.; Xiong, C.-L.; Yan, J.-C.; Mo, Y.-B.; Dou, S.-Q.; Wu, Z.-H.; Yang, R.-F. Citrus Pest Identification Model Based on Improved ShuffleNet. *Appl. Sci.* **2024**, *14*, 4437. [CrossRef]
22. Zhang, Y.; Xie, W.; Yu, X. Design and implementation of liveness detection system based on improved shufflenet V2. *Signal Image Video Process.* **2023**, *17*, 3035–3043. [CrossRef]
23. Paoletti, M.E.; Haut, J.M.; Pereira, N.S.; Plaza, J.; Plaza, A. Ghostnet for Hyperspectral Image Classification. *IEEE Trans. Geosci. Remote Sens.* **2021**, *59*, 10378–10393. [CrossRef]
24. Han, K.; Wang, Y.; Xu, C.; Guo, J.; Xu, C.; Wu, E.; Tian, Q. GhostNets on Heterogeneous Devices via Cheap Operations. *Int. J. Comput. Vis.* **2022**, *130*, 1050–1069. [CrossRef]
25. Lei, Y.; Pan, D.; Feng, Z.; Qian, J. Lightweight YOLOv5s Human Ear Recognition Based on MobileNetV3 and Ghostnet. *Appl. Sci.* **2023**, *13*, 6667. [CrossRef]
26. Li, H.; Li, J.; Wei, H.; Liu, Z.; Zhan, Z.; Ren, Q. Slim-neck by GSConv: A lightweight-design for real-time detector architectures. *J. Real-Time Image Process.* **2024**, *21*, 62. [CrossRef]
27. Jin, L.; Yu, Y.; Zhou, J.; Bai, D.; Lin, H.; Zhou, H. SWVR: A Lightweight Deep Learning Algorithm for Forest Fire Detection and Recognition. *Forests* **2024**, *15*, 204. [CrossRef]
28. Cui, M.; Duan, Y.; Pan, C.; Wang, J.; Liu, H. Optimization for Anchor-Free Object Detection via Scale-Independent GIoU Loss. *IEEE Geosci. Remote Sens. Lett.* **2023**, *20*, 1–5. [CrossRef]
29. Chen, J.; Zhu, J.; Li, Z.; Yang, X. YOLOv7-WFD: A Novel Convolutional Neural Network Model for Helmet Detection in High-Risk Workplaces. *IEEE Access* **2023**, *11*, 113580–113592. [CrossRef]
30. Liu, Q.; Huang, W.; Duan, X.; Wei, J.; Hu, T.; Yu, J.; Huang, J. DSW-YOLOv8n: A New Underwater Target Detection Algorithm Based on Improved YOLOv8n. *Electronics* **2023**, *12*, 3892. [CrossRef]
31. Zhang, T.; Wang, F.; Wang, W.; Zhao, Q.; Ning, W.; Wu, H. Research on Fire Smoke Detection Algorithm Based on Improved YOLOv8. *IEEE Access* **2024**, *12*, 117354–117362. [CrossRef]

Disclaimer/Publisher's Note: The statements, opinions and data contained in all publications are solely those of the individual author(s) and contributor(s) and not of MDPI and/or the editor(s). MDPI and/or the editor(s) disclaim responsibility for any injury to people or property resulting from any ideas, methods, instructions or products referred to in the content.

Article

Infrared and Visible Image Fusion Algorithm Based on Double-Domain Transform Filter and Contrast Transform Feature Extraction

Xu Ma [1,2,†], Tianqi Li [2,†], Jun Deng [1,*], Tong Li [2], Jiahao Li [2], Chi Chang [2], Rui Wang [2], Guoliang Li [2], Tianrui Qi [2] and Shuai Hao [2]

1. College of Safety Science and Engineering, Xi'an University of Science and Technology, Xi'an 710054, China; maxu@xust.edu.cn
2. College of Electrical and Control Engineering, Xi'an University of Science and Technology, Xi'an 710054, China; altq8792@163.com (T.L.); litong20221120@163.com (T.L.); 18406060425@stu.xust.edu.cn (J.L.); 23206223096@stu.xust.edu.cn (C.C.); 23206232146@stu.xust.edu.cn (R.W.); 23206232117@stu.xust.edu.cn (G.L.); 20406100226@stu.xust.edu.cn (T.Q.); haoxust@163.com (S.H.)
* Correspondence: dengj518@xust.edu.cn
† These authors contributed equally to this work.

Abstract: Current challenges in visible and infrared image fusion include color information distortion, texture detail loss, and target edge blur. To address these issues, a fusion algorithm based on double-domain transform filter and nonlinear contrast transform feature extraction (DDCTFuse) is proposed. First, for the problem of incomplete detail extraction that exists in the traditional transform domain image decomposition, an adaptive high-pass filter is proposed to decompose images into high-frequency and low-frequency portions. Second, in order to address the issue of fuzzy fusion target caused by contrast loss during the fusion process, a novel feature extraction algorithm is devised based on a novel nonlinear transform function. Finally, the fusion results are optimized and color-corrected by our proposed spatial-domain logical filter, in order to solve the color loss and edge blur generated in the fusion process. To validate the benefits of the proposed algorithm, nine classical algorithms are compared on the LLVIP, MSRS, INO, and Roadscene datasets. The results of these experiments indicate that the proposed fusion algorithm exhibits distinct targets, provides comprehensive scene information, and offers significant image contrast.

Keywords: high-pass filter; logical filter; feature extraction; nonlinear contrast transform function; color correction

1. Introduction

In the field of image fusion, infrared and visible image fusion plays a crucial role and finds extensive applications [1,2]. Within the military domain [3,4], it facilitates target identification, night vision capabilities, and navigation systems. In terms of security applications [5], it enhances nighttime or complex-environment surveillance by providing clearer images to aid in detecting abnormal behavior. The medical sector [6] benefits from its usage for disease diagnosis and treatment monitoring purposes. By employing a specific fusion method to combine infrared and visible images effectively, this technique ensures that the resulting fused image encompasses feature information from both sources while presenting a more comprehensive and accurate depiction [7]. So far, the most widely used method in the field of image fusion is based on a pixel-level approach. This method is broadly categorized into the following four types: spatial-domain [8], transform-domain [9], low-rank matrix [2], and bionic algorithms [10].

Spatial-domain image fusion algorithms that are widely used include but are not limited to weighted average [11], contrast transform [12], and logical filtering algorithms [13].

Spatial-domain image fusion algorithms operate directly on the pixel values of the image, which is simple, intuitive, and easy to understand and has better real-time performance.

Transform-domain image fusion algorithms convert images from their time domain to the frequency domain. They fall into two main groups: frequency-domain filters and multiscale transform techniques. Frequency-domain filtering methods excel in feature extraction, capturing more image details and texture features compared to spatial-domain fusion algorithms. Multiscale transformation is widely recognized as an effective technique for achieving image fusion. The image fusion algorithm based on a multiscale transform can be divided into three major steps: multiscale decomposition, fusion rules, and inverse multiscale transform [14]. Currently, widely used multiscale transform fusion algorithms can be categorized into two main groups according to the transform rules: pyramid transforms and wavelet transforms. A pyramid transform contains contrast pyramids, Laplace pyramid transform, and so on. In the field of image fusion based on the wavelet transform, Pu et al. [15] proposed an image fusion algorithm based on contrast transformation and wavelet decomposition. In addition, Li et al. [16] proposed an image fusion method combined with morphological image enhancement and a dual-tree complex wavelet. This method was a good solution for images with ringing, incomplete scene information, low contrast, and Gibbs artifacts. Li et al. [17] introduced an image fusion scheme based on NSCT and low-level visual features. Moreover, Tan et al. [18] proposed an image fusion via NSST and PCNN [19] in the multiscale morphological gradient (MSMG) domain and explored a new fusion method in the MSMG domain. On the basis of NSST, Li et al. [20] proposed the LSWT-NSST image fusion algorithm, which effectively solved edge blurring and Gibbs phenomena in the traditional wavelet transform algorithm and the loss of subtle features in the NSST.

Low-rank matrix algorithms can be broadly classified into three groups depending on the techniques employed: matrix decomposition algorithms, matrix representation algorithms, and approaches utilizing low-rank kernel approximation. Matrix decomposition methods work by stacking multiple images into a matrix and then performing a low-rank decomposition of that matrix. Suryanarayana et al. [21] designed the multiple degradation skilled network for infrared and visible image fusion based on a multi-resolution SVD update, effectively improving the resolution of the fused image. Principal Component Analysis (PCA) [22] has desirable denoising effect. It has no parameter limit in the fusion process, and the calculation is small. Wang et al. [23] presented an improved image fusion method based on NSCT and accelerated Non-negative Matrix Factorization (NMF) and obtained better fusion results than with PCA. Low-rank representation methods are based on the assumption that the information of an image can be represented by a low-rank matrix. Image fusion can be achieved by decomposing each image into a combination of a low-rank matrix and a sparse matrix and then performing a weighted summation on the low-rank part. By combining a Low Rank Representation (LRR) and a Sparse Representation (SR), Wang et al. [24] proposed a low-rank sparse representation (LSRS) to provide new theoretical knowledge for image fusion. Low-rank kernel approximation-based methods perform image fusion by transforming the image convolution operation into a low-rank influence. Abdolali et al. [25] introduced a multiscale decomposition in the low-rank approximation in their paper and provided a theoretical basis for a Multiscale Low-Rank Approximation (MLRA) which was then applied to image processing. Low-rank matrix algorithms can have disadvantages such as high computational complexity, poor noise immunity, and the need for prior knowledge in some cases. When applying these algorithms, it is crucial to consider their limitations. To achieve optimal fusion results, we must carefully select appropriate parameters and make adjustments to the algorithm design based on the specific situation.

An important branch of bionic algorithms is the deep learning algorithms based on neural networks, which typically enforce desired distributional properties in the fused images by constructing an objective function. Convolutional Neural Networks (CNNs) [26] have gained popularity in image fusion due to their powerful feature extraction capabilities.

Deep learning approaches usually adopt an end-to-end strategy for fusion, minimizing the need for preprocessing, parameter tuning, and post-processing. However, there are limited training data available for deep learning methods in image fusion compared to tasks like target recognition and detection, which affects the performance of deep learning models.

Currently, infrared and colorful visible images are a popular research topic with diverse applications in daily life. However, two issues arise during the fusion process: a loss of texture details due to color information loss and edge blurring of the infrared target. To address these problems, we propose a new algorithm called DDCTFuse that is based on a double-domain transform filter and nonlinear transform feature extraction. Our specific contributions are listed below:

(1) In order to improve the quality and flexibility of images in the frequency-domain decomposition, an adaptive high-pass filtering algorithm is proposed and used in the frequency-domain decomposition. Meanwhile, the high-frequency texture is denoised and the edge features are preserved by introducing bilateral filtering.

(2) Aiming at the problem that the infrared target of fusion image is not obvious, in this paper, a novel nonlinear element transform function is designed, and a new feature extraction algorithm for infrared targets is proposed. The algorithm is very sensitive to the high-energy part of infrared images and can significantly enhance and extract infrared targets.

(3) A novel spatial-domain logical filter image optimization strategy is proposed, which effectively decreases "virtual shadow" (artifacts) and preserves fine details caused by the Gibbs phenomenon (ringing effect) in the frequency-domain fusion process.

The remaining sections of the paper are structured as follows: In Section 2, we provide a concise overview of the bilateral filter and establish the theoretical foundation for the subsequent sections. In Section 3, we describe the process and principle of DDCTFuse in detail. Experimental details, parameter settings, and result analysis are provided in Section 4. Finally, results are discussed in Section 5 and summarized in Section 6.

2. Bilateral Filter

The bilateral filter (BF) is a well-known edge-preserving filter and a classical nonlinear spatial filtering method. The bilateral filter integrates the effects of the spatial and value domains on the filtering generation; thus, it can achieve the effect of noise reduction and smoothing and is a good edge-preserving filter. The bilateral filter was employed in this paper to effectively attenuate high-frequency details while preserving the intricate texture, thereby achieving high-frequency denoising.

A bilateral filter is realized by a weighted average method based on a Gaussian distribution. The principle of the bilateral filter implementation is illustrated by solving specific pixel values in the image as an example. The output pixel value g of the image at (i,j) after bilateral filtering depends on the weighted combination of pixel values f in the neighborhood.

$$P(i,j) = \frac{\sum_{k,l} F(k,l) E(i,j,k,l)}{\sum_{k,l} E(i,j,k,l)} \qquad (1)$$

where k and l denote the position of the neighboring pixel, and the weight coefficient $E(i,j,k,l)$ is the product of the spatial-domain kernel d and the value-domain kernel r. The spatial-domain kernel d is the Euclidean distance between the current point and the center point calculated based on the Gaussian function.

$$Y(i,j,k,l) = \exp\left(-\frac{(i-k)^2 + (j-l)^2}{2\sigma_Y^2}\right) \qquad (2)$$

The value-domain kernel r is the absolute value of the difference between the current point and the center pixel value based on a Gaussian function.

$$T(i,j,k,l) = \exp\left(-\frac{\|f(i,j) - f(k,l)\|^2}{2\sigma_T^2}\right) \quad (3)$$

From the formulae for the space-domain kernel and the value-domain kernel, the weight coefficients can be calculated as:

$$E(i,j,k,l) = \exp\left(-\frac{(i-k)^2 + (j-l)^2}{2\sigma_Y^2} - \frac{\|f(i,j) - f(k,l)\|^2}{2\sigma_T^2}\right) \quad (4)$$

In order to more intuitively reflect the effect of the algorithm, the specific details of the enhanced output results are shown in Figure 1.

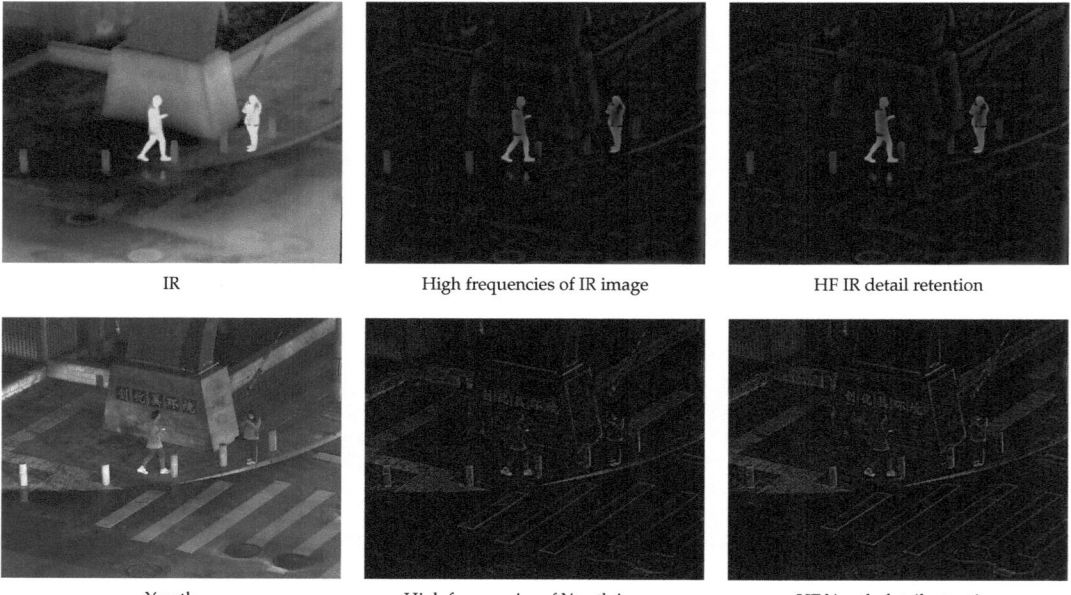

Figure 1. BF detail retention.

3. The Proposed DDCTFuse Method

In this section, we provide a detailed introduction of DDCTFuse, as shown in Figure 2. Also, the mathematical principles of DDCTFuse are presented in this section.

The YCbCr color space transform has the advantages of being better at processing image brightness and color information, reducing data redundancy, and improving compression efficiency. As a result, the Y-channel information of the RGB visible image was extracted using the YCbCr transform. Right after that, the Y-channel image (Y) and infrared image (IR) were decomposed into high-frequency and low-frequency components through the adaptive high-pass filter. Then, the extracted high-frequency detail features (including noise) were processed using BF, preserving the detail texture while reducing the noise. Meanwhile, the nonlinear contrast transform feature extraction was performed on the IR image to obtain a feature map, which was used to guide the fusion of the Y-channel image and the high-frequency part of the IR image. The low-frequency part was fused by wavelet weighting to obtain the final base layer. Afterwards, a spatial logic filter optimization strategy was devised to enhance the infrared target while preserving the color and texture information of the source image, effectively optimizing the quality of the fused image and obtaining the final result. Finally, an inverse YCbCr transform generated the final fusion result.

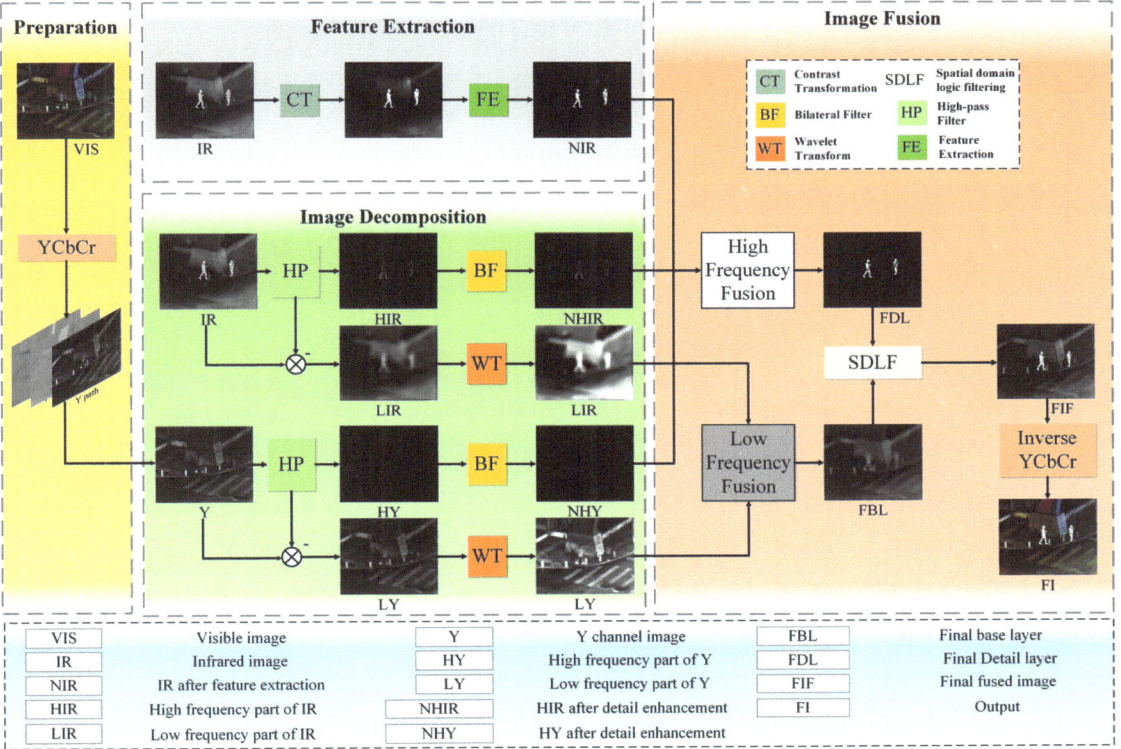

Figure 2. Schematic of the proposed DDCTFuse method: the gray dotted line box illustrates the specific algorithmic steps, the blue box presents the full image name, the black dotted line box presents the complete names of distinct algorithms within the fusion strategy, facilitating readability.

3.1. Adaptive High Pass Filter

The high-frequency component of an image represents sudden changes in intensity, which can correspond to edges or noise, and often both. The low-frequency component represents the portion of the image that undergoes gradual changes, specifically referring to the contour information within the image.

In traditional Butterworth filtering, the cutoff frequency is usually set to a certain value and applied to all images. This is not conducive to the extraction of high-frequency details, which often leads to the incomplete extraction of high-frequency details or high-frequency information containing low-frequency components, which reduces the robustness of the algorithm.

In order to solve the above problems, we designed an adaptive high-pass filter and dynamically adjusted the cutoff frequencies. The mean frequency of the image can be obtained by averaging the amplitude. We took the mean frequency as the cutoff frequency of the high-pass filter to preserve more detail in the image. The specific mathematical expression is given below.

First, we performed a domain transformation of the image. The Fourier expression of the image is discontinuous and its discrete expression is:

$$F(u,v) = \frac{1}{MN} \sum_{x=0}^{M-1} \sum_{y=0}^{N-1} f(x,y) e^{-j2\pi(\frac{ux}{M} + \frac{vy}{N})} \qquad (5)$$

where $u = 0, 1, 2, \ldots, M-1$; $v = 0, 1, 2, \ldots, N-1$; $f(x,y)$ denotes the grayscale value of the point; M and N denote the length and width of the image; $F(u,v)$ denotes the Fourier transform result of the image.

Second, the cutoff frequency was calculated. In order to find the mean frequency of the image, the Fourier transform of the image was transformed into a polar coordinate representation.

$$F(u,v) = |F(u,v)| e^{-j\phi(u,v)} \qquad (6)$$

where $|F(u,v)|$ denotes the magnitude and ϕ denotes the phase angle.

The expression for the magnitude is:

$$|F(u,v)| = [R(u,v)^2 + I(u,v)^2]^{\frac{1}{2}} \qquad (7)$$

where $R(u,v)$ and $I(u,v)$ denote the real and imaginary parts of $F(u,v)$.

Then, the frequency coordinates corresponding to the average amplitude were calculated, and their mathematical expressions are shown below:

$$A_0 = \frac{1}{M*N} \sum_{u=0}^{M-1} \sum_{v=0}^{N-1} |F(u,v)| \qquad (8)$$

$$(u_0, v_0) = \arg\min_{(u,v)} |F(u,v) - A_0| \qquad (9)$$

where A_0 is the average amplitude of the image, and (u_0, v_0) is the frequency coordinate that minimizes the difference between the amplitude spectrum $F(u,v)$ and A_0.

The frequency values corresponding to the frequency coordinates (u_0, v_0) can be obtained by the following formula:

$$f_{u0} = u_0 \cdot \Delta f_u = \frac{B_h}{M} \qquad (10)$$

$$f_{v0} = v_0 \cdot \Delta f_v = \frac{B_v}{N} \qquad (11)$$

$$F_0 = \frac{(f_{v0} + f_{u0})}{2} \qquad (12)$$

where f_{v0} and f_{u0} represent the horizontal and vertical components of the frequency domain, respectively, B_h and B_v represent horizontal and vertical bandwidths, and F_0 is the average frequency.

According to the mathematical formula of the Butterworth high-pass filter, we propose the following formula for the adaptive high-pass filter introduced in this paper:

$$H(u,v) = \frac{1}{1 + [F_0/D(u,v)]^{2n}} \qquad (13)$$

Third, a frequency-domain decomposition was performed. The low-frequency part was obtained by performing a two-dimensional Fourier inverse transform of H into the time domain (h) and subtracting h (the high-frequency part) from f.

$$h(x,y) = \frac{1}{M*N} \sum_{u=0}^{M-1} \sum_{v=0}^{N-1} H(u,v) e^{-j2\pi(\frac{ux}{M} + \frac{vy}{N})} \tag{14}$$

$$l(x,y) = \sum_{x=0}^{M-1} \sum_{y=0}^{N-1} [f(x,y) - h(x,y)] \tag{15}$$

where $h(x,y)$ denotes the high-frequency part pixel value, and $l(x,y)$ denotes the low-frequency part pixel value.

3.2. Contrast Transform

We propose a new nonlinear element transform function to obtain the image contrast conversion coefficient CT and weight the image pixel value to transform the image contrast.

First, the optimal threshold "G" was obtained by the Nobuyuki Otsu method. A summary of the pixels in the light (B) and dark (D) parts of the image was obtained by comparing and counting the entire image pixel by pixel with "G". We defined a luminance factor "a" to represent the light-to-dark ratio of the image. It is worth noting that in order to clarify the following statements, we define some variables: $OTSU(\bullet)$ indicates that the optimal threshold is obtained by the Nobuyuki Otsu method, and $poll[\bullet]$ indicates that images are polled, compared, and counted.

We calculated the luminance factor "a" as follows:

$$G = OTSU(F_{in}) \tag{16}$$

$$\begin{cases} B = poll[F_{in}(x,y) \geq G] \\ D = poll[F_{in}(x,y) < G] \end{cases} \tag{17}$$

$$a = \frac{B}{D} \tag{18}$$

where F_{in} is the input figure.

Second, the contrast conversion coefficient CT was obtained by designing a new nonlinear element function. The image was weighted to obtain the pixel value $F_{trans}(x,y)$ and the image F_{trans} after the final transform. The specific mathematical expression is as follows:

$$CT = \frac{R*(1/a)}{T - F_{in}(x,y) + \varepsilon} \tag{19}$$

$$F_{trans}(x,y) = CT * F_{in}(x,y) \tag{20}$$

where $F_{in}(x,y)$ is the pixel value of the image in a certain spatial coordinate; R is the average pixel value of the input figure; T is the maximum pixel value of the input figure; ε is an infinitesimal quantity to prevent the denominator from being zero.

Finally, the two image matrices F_{trans} and F_{in} were transformed by a linear difference to obtained the target feature F_{tf}, where a is the luminance factor:

$$F_{tf} = \begin{cases} F_{trans} - F_{in}*a, a \leq 1 \\ F_{trans} - F_{in}/a, a > 1 \end{cases} \tag{21}$$

The specific feature extraction process is shown in Figure 3.

| IR | Contrast transform | Feature Extraction |

Figure 3. Contrast transform feature extraction process.

3.3. Spatial-Domain Logic Filter

The pixel values of the fused image were corrected by a logical filter to reduce the artifacts as well as the color loss due to the Gibbs effect during the transform domain filtering process.

First, the final base layer and the final detail layer were fused by linear superposition and then converted into a matrix. Multiple iterations were subsequently conducted for pixel-wise comparisons of grayscale values. Ultimately, the optimized outcome was achieved by assigning appropriate grayscale values to the resultant image. The specific process is shown in Figure 4.

Below, we focus on the mathematical principles of the logical filter. Two 3×3 sub-blocks (A and B) were extracted from the initial fusion image (F) and the Y-channel image (Y), respectively. A difference matrix (C) was generated by the difference calculation for these two sub-blocks, and we calculated the mean difference (C_m) in that region. Next, C_m was compared with the difference matrix, element-by-element, to generate a new matrix (I_{sub}). The following is the mathematical analysis.

First, we obtained the difference matrix (C).

$$\begin{bmatrix} c_1 & c_2 & c_3 \\ c_4 & c_5 & c_6 \\ c_7 & c_8 & c_9 \end{bmatrix} = \left| \begin{bmatrix} a_1 & a_2 & a_3 \\ a_4 & a_5 & a_6 \\ a_7 & a_8 & a_9 \end{bmatrix} - \begin{bmatrix} b_1 & b_2 & b_3 \\ b_4 & b_5 & b_6 \\ b_7 & b_8 & b_9 \end{bmatrix} \right| \tag{22}$$

Then, we calculated the mean of the matrix C and defined it as the mean difference (C_m) of the two sub-blocks A and B

$$C_m = \frac{1}{D^2} \sum_{n=0}^{D^2} (c_n) \tag{23}$$

The next step was to poll C_m to compare the elements in the matrix C and make a logical judgment as shown in Formula (24) to adjust the elements in the matrix C. We named the optimized C matrix I_{sub} and defined the above method as logical filtering.

$$\begin{cases} c_n = a_n, if c_n \leq C_m \\ c_n = b_n, if c_n > C_m \end{cases} \tag{24}$$

Finally, the initial fusion image F was filtered by the logical filter with a step size of 1, and the optimized image I_{Fused} was obtained by reconstructing the matrix I_{sub}.

Figure 5 shows in detail the part of the optimization algorithm that relates to the logical filtering of fusion results. It can be seen that the detailed texture of I_{Fused} is richer than F, and the Gibbs phenomenon is eliminated to a great extent without losing the target edge.

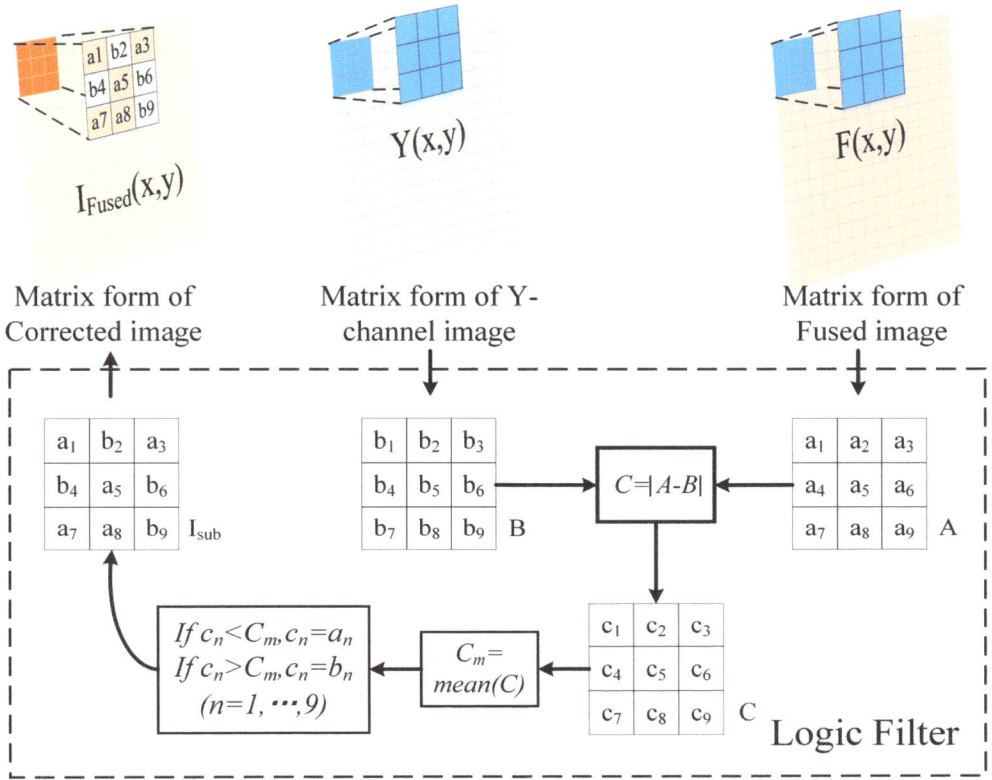

Figure 4. Logic filtering process: the elements of F, denoted as $F(x,y)$, represents the pixel value of F. $a_n, (n = 1, ..., 9)$ denotes the pixel value in sub−block A. The pixel value of the Y−channel image is represented by $Y(x,y)$. $b_n, (n = 1, ..., 9)$ denotes the pixel value in sub−block B. C represents the difference matrix.

(a) Unoptimized (b) Optimized

Figure 5. Comparison of logic filter optimization results. The infrared target edge details have been marked in red boxes and the background details have been marked in green boxes.

To improve understanding, Algorithm 1 provides a pseudocode depiction of the proposed DDCTFuse method.

Algorithm 1 DDCTFuse

Input: *Infrared image* (IR) and *visible image* (VIS);
    Step 1: Color-space transformations
        Convert RGB and VIS to $YCbCr$
        $(Y, Cb, Cr) = YCbCrtransformation(VIS)$,
    Step 2: Image decomposition
        do: IR and Y are decomposed into high-frequency parts (HIR, HY) and a low-frequency parts (LIR, LY) by the adaptive high-pass filtering.
        $(HIR, LIR) = BWfilter(IR)$,
        $(HY, LY) = BWfilter(Y)$,
        then: Image detail retention of the high-frequency parts (HIR, HY) is performed using BF (bilateral filtering) to obtain $NHIR$ and NHY.
        $(NHIR) = BFfilter(IR)$,
        $(NHY) = BFfilter(Y)$,
    Step 3: Contrast transformation
        The contrast-transformed features are extracted from IR by the proposed contrast transform function to obtain the feature map NIR.
        $NIR = ContrastTransform(IR)$,
    Step 4: High-frequency fusion method
        NIR is utilized to guide the fusion of the high-frequency parts $(NHIR, NHY)$ to obtain the final detail layer (FDL).
        $FDL = concat(NHIR, NHY, NIR)$,
    Step 5: Low-frequency fusion method
        The low-frequency parts (LIR, LY) are wavelet-decomposed and fused with a weighted average using the wavelet weighted-average fusion strategy to obtain the final base layer (FBL).
        $FBL = wavelet(LIR, LY)$,
    Step 6: Image fusion
        FBL and FDL are fused to obtain the fusion result (IF).
        $IF = concat(FBL, FDL)$,
    Step 7: Optimization of fusion results
        The fused image is optimized using the designed logical filtering optimization strategy based on a spatial-domain grayscale comparison to obtain the final result (FIF).
        $FIF = Optimization(IF)$,
    Step 8: Inverse color-space conversion
        The fused image (FI) is obtained by using the inverse YCbCr transform on FIF.
        $FI = InverseYCbCr(FIF, Cb, Cr)$,
Output: Fused image (FI);

---

4. Experiments and Analysis

4.1. Experimental Setup

The DDCTFuse method proposed in this paper was built in the MATLAB R2023a software environment, and all experiments were performed on the same hardware platform (NVIDIA GeForce RTX 4060Ti GPU, i5-13490F CPU) and software platform (Window11 operating system). In order to verify the superiority of DDCTFuse, nine infrared and visible image fusion methods were selected for comparison, including ADF [27], LatLRR [28], DenseFuse [29], IFCNN [30], GFF, DIVFusion [31], NestFuse [32], RFN-Nest [33], and SeAFusion [34].

The test datasets were LLVIP [35], MSRS [36], INO, and Roadscene [37]. LLVIP contains 12,025 groups of IR and visible images. MSRS contains 361 groups of IR and visible images. INO contains 55 groups of IR and visible images. EN, AG, MI, VIF, Nabf, and FMI were used as objective evaluation indexes of the fused image quality. To demonstrate the advantages of our adaptive high-pass filtering, based on the contrast transformation of nonlinear unit function, spatial-domain logic filtering, and low-frequency partial fusion strategy, we devised three sets of ablation experiments. In addition, we conducted real-time testing experiments to demonstrate the important development prospects of DDCTFuse.

4.2. Subjective Evaluation

To illustrate the advantages of DDCTFuse, we compared it with nine different classical algorithms on three different datasets. In order to show some of the image details, the infrared target part in the following figure is marked with a zoomed-in red box, and the background detail information is marked with a zoomed-in green box.

In Figure 6, SeAFusion, Nestfuse, and IFCNN have a strong ability to extract the infrared energy information in the source images, but the road texture cannot be accurately extracted in low-light conditions. Although the results of GFF retain some background information of the source visible image, the infrared target has obvious Gibbs artifacts and edge loss. DIVFusion, ADF, DenseFuse, RFN-Nest, and LatLRR demonstrate a heightened focus on global information, which leads to the inconspicuous infrared target. In comparison, our algorithm's results possess a very significant and high-quality infrared target while preserving the source's visible background details as well as color information.

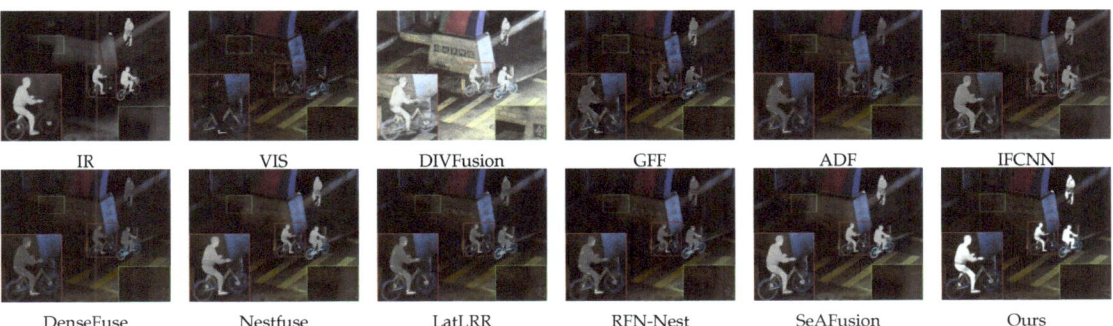

Figure 6. Comparison results on "010370" in the LLVIP dataset.

As shown in Figure 7, ADF, DenseFuse, GFF, LatLRR, and RFN-Nest give a parallel level of attention to information in both source images, which creates a dim fused image. IFCNN, DIVFusion, and Nestfuse put more emphasis on preserving the color information of the visible image, and clearly extract the details of the tree and the house, which leads to some loss of the infrared target and causes ringing artifacts. SeAFusion and DDCTFuse have distinct IR targets and retain the background information of the source visible image to a large extent, and our proposed method has a subtle advantage in color preservation, as shown in the green-box enlargement section.

In Figure 8, GFF, ADF, IFCNN, DenseFuse, and LatLRR render enough attention to the global information of both source images, which result in the fused image possessing segmental background information of the infrared image. DIVFusion and RFN-Nest can well extract the contrast information of the visible image; in other words, it leads to the loss of a fraction of the infrared energy information. Nestfuse, SeAFusion, and DDCTFuse have a natural fusion effect, and the infrared target is prominent; moreover, the texture of the tree is relatively clear.

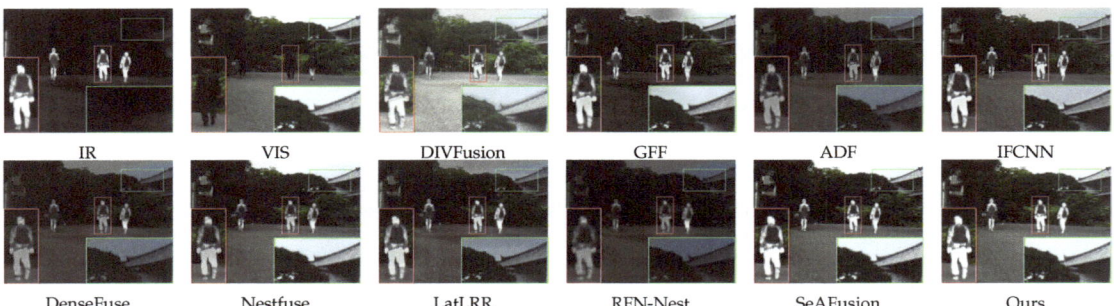

Figure 7. Comparison results on "00634D" in the MSRS dataset.

In Figure 9, since the visible image itself is blurred and of low quality, it results in RFN-Nest, DIVFusion, and DenseFuse focusing on the global features of the visible image, therefore presenting a misty fused image. GFF, IFCNN, ADF, Nestfuse, and SeAFusion have a strong ability to extract infrared energy information. However, they are unable to pay close attention to the edge information, which leads to artifacts. LatLRR and our method possess clear fusion results and significant infrared targets, and even the branches have rich detailed information.

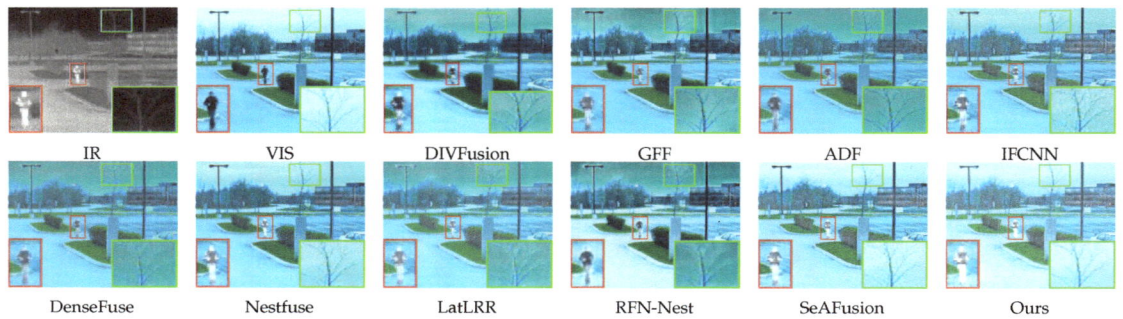

Figure 8. Comparison results on "44" in the INO dataset.

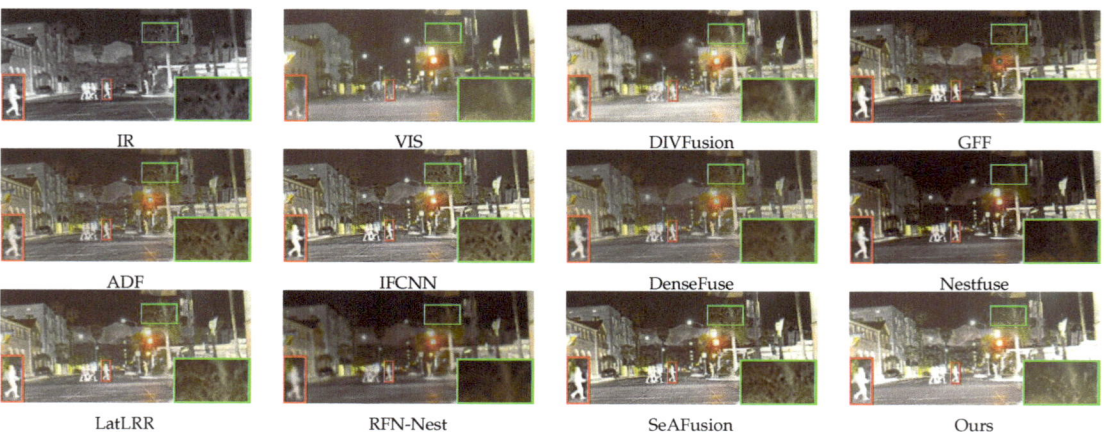

Figure 9. Comparison results on "FLIR-09016" in the Roadscene dataset.

4.3. Objective Evaluation

A subjective evaluation has a certain degree of one-sidedness and is susceptible to human factors. Therefore, we chose the following objective evaluation metrics: entropy (EN), average gradient (AG), mutual information (MI), No-reference assessment based on blur and noise factors (Nabf), Feature Mutual Information (FMI), and visual information fidelity (VIF) for our analysis. In total, 80 images from the MSRS dataset, 100 images from the LLVIP dataset, and all images in the Roadscene and INO datasets were selected for testing. In this case, except for Nabf, which takes a smaller value, a larger value is better for the other five objective evaluation indicators. The definitions of each metric are as follows:

(1) The information entropy (EN) serves as an objective assessment metric for quantifying the amount of information present in an image. In this context, "a" represents the grayscale value, while $P(a)$ signifies the probability distribution of gray levels. A higher EN value indicates a greater abundance of information and superior quality in the fused image.

$$H(A) = -\sum_a P_A(a) \log p_A(a) \tag{25}$$

(2) The clarity of the fused image can be evaluated using the average gradient (AG), where a higher average gradient indicates superior image clarity and fusion quality. The formula for computing AG is as follows:

$$AG = \frac{1}{(M-1)(N-1)} \sum_{i=1}^{M-1} \sum_{i=1}^{N-1} \sqrt{\frac{(H(i+1,j) - H(i,j))^2 + (H(i,j+1) - H(i,j))^2}{2}} \tag{26}$$

where H is the composite image, with M and N representing its dimensions in terms of height and width, respectively.

(3) The mutual information (MI) serves as a quantification of the degree of similarity between two images, indicating the extent to which information from the original image is preserved in the fused image. A higher mutual information signifies superior fusion quality by retaining more information from the source image. The computation of mutual information involves considering both individual images' entropy $H(A)$ and their joint entropy $H(A, B)$.

$$MI(A, B) = H(A) + H(B) - H(A, B) \tag{27}$$

(4) The complete designation of Nabf is "No-reference evaluation relying on factors of blur and noise", indicating its function as a no-reference method for assessing image quality based on the presence of blur and noise. It accurately determines the degree of image blurriness and noise without requiring a reference image for comparison.

(5) The fidelity of information is commonly assessed using the visual information fidelity (VIF) metric, which serves as a measure for evaluating the subjective visualization quality in images. A higher VIF value indicates a superior visual representation. The calculation is executed through a four-step process, accompanied by the following simplified formula:

$$VIF = \frac{VID}{VIND} \tag{28}$$

where the acronyms VID and $VIND$ denote the visual information derived from the composite image generated by the input image.

(6) Feature Mutual Information (FMI) is a method for measuring the similarity between two images, which is based on feature information theory and mutual information theory. The similarity between two images is quantified by a metric that measures the occurrence of shared pairs of feature points. The formula provided below presents a straightforward approach for evaluating this aspect:

$$FMI_F^{AB} = \frac{I_{FA}}{H_F + H_A} + \frac{I_{FB}}{H_F + H_B} \tag{29}$$

where F denotes the fused image, and A, B are the source images; I_{FA} and I_{Fb} are the feature information of A and B contained in F represented through MI; the entropy values based on the histograms of images A, B, and F are represented by HA, HB, and HF, respectively.

In Table 1, the average values of 80 randomly selected sets of image evaluation metrics in the MSRS dataset are shown, where the optimal values are marked in bold, the second best in red, and the third in blue.

Table 1. Objective evaluation metrics of 9 comparison algorithms and our proposed algorithm on the MSRS dataset.

Algorithm	EN	AG	MI	VIF	Nabf	FMI
ADF	5.9054	2.6496	2.4144	0.7080	0.0605	0.9106
DenseFuse	5.8657	2.1403	2.4066	0.6675	0.0504	0.9193
DIVFusion	7.4803	4.5438	2.4721	0.7668	0.3079	0.8799
GFF	6.2610	3.3391	2.2872	0.8714	0.0773	0.9323
IFCNN	6.3944	4.0352	2.6651	0.8124	0.2364	0.9191
LatLRR	6.2447	2.7243	2.2384	0.7826	0.0759	0.9200
NestFuse	6.0739	3.3290	3.4932	0.8679	0.1438	0.9261
RFN-Nest	5.4312	2.0016	2.1027	0.6012	0.0496	0.9189
SeAFusion	6.4042	3.8455	3.4686	0.9401	0.2155	0.9263
Ours	6.5959	3.3702	5.8577	0.9695	0.0460	0.9347

From the objective evaluation metrics in the above table, it can be seen that the proposed algorithm occupies four optimal solutions and one suboptimal result, which proves that the fused image contains more detail information and has high fusion quality and great contrast.

We selected one hundred images from the LLVIP dataset for testing, and our method performed well, with optimal values for AG, MI, Nabf, and FMI, as shown in Table 2. This substantiates DDCTFuse's advantages and underscores its superior robustness.

Table 2. Objective evaluation metrics of 9 comparison algorithms and our proposed algorithm on the LLVIP dataset.

Algorithm	EN	AG	MI	VIF	Nabf	FMI
ADF	6.3888	1.7793	2.6872	0.8978	0.1069	0.9403
DenseFuse	6.3785	1.5676	2.7953	0.8989	0.08522	0.9460
DIVFsuion	7.4952	4.5587	2.5919	0.7943	0.3264	0.9033
GFF	6.3560	2.1034	2.6164	1.1334	0.1186	0.9411
IFCNN	6.6042	2.1479	3.0449	0.9823	0.1540	0.9396
LatLRR	6.4300	1.4789	2.5881	0.8466	0.08545	0.9363
NestFuse	6.7731	1.6542	3.5813	1.0986	0.0842	0.9413
RFN-Nest	6.5803	1.3783	2.8061	0.9908	0.08539	0.9324
SeAFusion	6.5947	2.0777	3.6510	1.1364	0.1380	0.9441
Ours	6.6521	2.1506	4.2829	1.0526	0.0834	0.9671

The Roadscene dataset was tested, and all 221 images were examined. The comparison results between DDCTFuse and nine other image fusion algorithms on the Roadscene dataset are presented in Table 3, where MI, VIF, FMI, and Nabf had optimal values. The results of the objective evaluation metrics demonstrate that DDCTFuse effectively preserves the intricate details of the source images, exhibits strong consistency with subjective perception, and yields high-quality fused images.

Table 3. Objective evaluation metrics of 9 comparison algorithms and our proposed algorithm on the Roadscene dataset.

Algorithm	EN	AG	MI	VIF	Nabf	FMI
ADF	7.0962	6.0353	1.0238	0.0802	0.7495	0.3788
DenseFuse	6.9735	3.7808	1.1083	0.0857	0.7708	0.2779
DIVFsuion	7.5506	4.7561	2.8942	0.7129	0.8429	0.1427
GFF	7.5728	5.0169	3.3935	0.0963	0.8024	0.1764
IFCNN	7.2164	6.1860	1.0298	0.0789	0.7627	0.3937
LatLRR	7.0560	4.0818	1.1083	0.0883	0.7707	0.2871
NestFuse	7.5735	5.6342	1.2769	0.0916	0.7734	0.3490
RFN-Nest	7.4774	3.7637	1.3288	0.1003	0.7914	0.2785
SeAFusion	7.4850	7.2913	1.2021	0.0863	0.7642	0.4269
ours	6.9131	5.9633	**5.6008**	**0.7763**	**0.8695**	0.1331

Shown in Table 4 are the test results of all algorithms on the INO dataset. Each algorithm had its own advantages. DDCTFuse obtained the best MI and Nabf values, which indicates that the fused image can better retain the information of the source image.

Table 4. Objective evaluation metrics of 9 comparison algorithms and our proposed algorithm on the INO Dataset.

Algorithm	EN	AG	MI	VIF	Nabf	FMI
ADF	6.8171	5.2523	2.9866	0.7244	0.8301	**0.0655**
DenseFuse	6.7553	4.2104	2.9034	0.7126	0.8557	0.0842
DIVFsuion	7.4305	5.7465	3.2211	0.8111	0.8468	0.1529
GFF	7.1982	6.5936	3.6897	0.9121	0.8847	0.0773
IFCNN	6.8974	7.5545	2.4015	0.7096	0.8599	0.2811
LatLRR	6.8863	5.0988	2.6903	0.6749	0.8476	0.1293
NestFuse	7.0922	6.6856	2.8999	0.8155	0.8549	0.1966
RFN-Nest	7.3883	5.8763	3.2622	0.7875	0.8605	0.1478
SeAFusion	7.1964	8.1769	2.7893	0.7744	0.8671	0.3223
Ours	6.8644	7.0469	**4.6567**	0.8726	**0.8906**	0.1873

4.4. Ablation Experiments

This section aims to further showcase the advantages of our adaptive high-pass filter, contrast transform feature extraction, and logical filtering optimization algorithms. Due to space constraints, we conducted three sets of ablation experiments on the LLVIP and MSRS datasets and analyzed the results.

The first set of ablation experiments aimed to highlight the benefits of the adaptive high-pass filter in the frequency-domain image decomposition. We replaced the adaptive high-pass filter with two commonly used high-pass filters: a traditional Gaussian high-pass filter and an ideal filter. The specific results are presented in Figures 10–12.

The experimental results clearly demonstrated the effectiveness of our proposed adaptive high-pass filter in accurately extracting both high- and low-frequency components from images. Compared to other prevalent high-pass filters, our innovative approach generated a more comprehensive and distinctive depiction of the target's high-frequency details, thereby enhancing overall accuracy and effectiveness in image processing tasks.

The adaptability of our proposed high-pass filter was particularly crucial when dealing with images that had significant variations in their characteristics. By dynamically adjusting the filter parameters based on each image's specific features, our approach ensured accuracy in extracting both high- and low-frequency components, even when faced with complex or challenging image content.

Furthermore, our adaptive high-pass filter excelled at preserving the integrity of the target image's important high-frequency details, which are essential for accurate interpretation and analysis of visual information.

In order to prove the outstanding performance of contrast transform feature extraction and logical filtering optimization in nonlinear fusion in the spatial-domain, a second set of ablation experiments was designed as follows: deletion of the contrast transform feature extraction, deletion of the logical filtering optimization in the spatial-domain, deletion of both features. The specific results are presented in Figures 13 and 14.

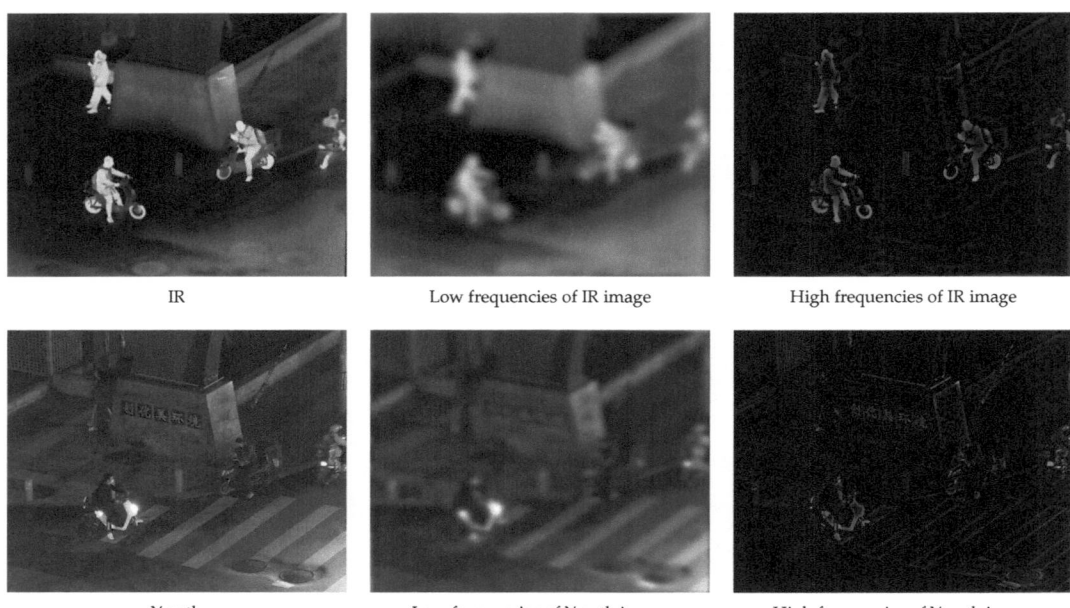

Figure 10. Adaptive high-pass filtering frequency-domain decomposition results.

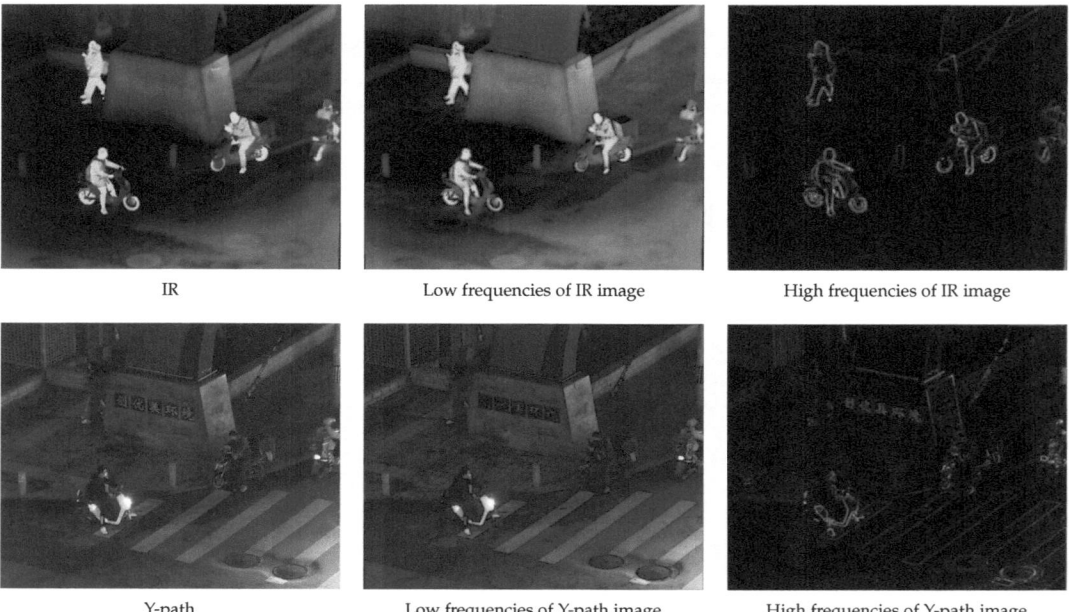

Figure 11. Gaussian high-pass filter frequency-domain decomposition results.

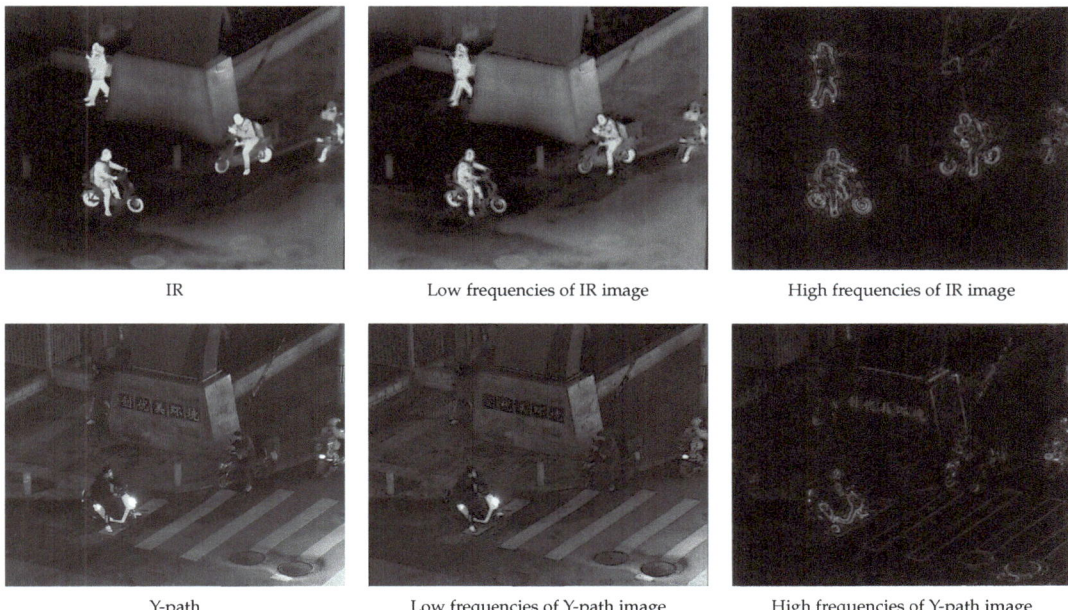

Figure 12. Ideal high-pass filter frequency-domain decomposition result.

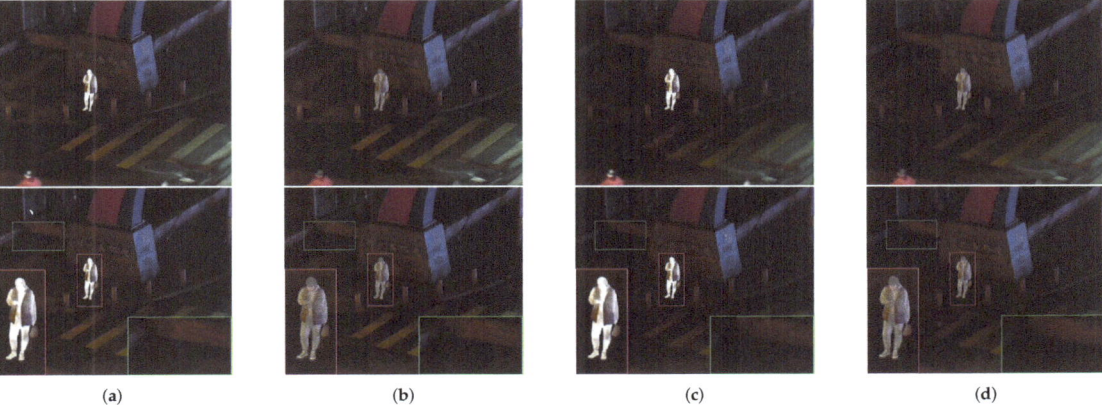

Figure 13. Ablation experiments on "010037" in the LLVIP dataset. The infrared target edge details have been marked in red boxes and the background details have been marked in green boxes.

In the set of figures presented above, we have four distinct outputs that serve to illustrate the effectiveness of our proposed method for fusing infrared images.

The outputs of DDCTFuse in column (a) serve as a reference for evaluating the effectiveness of the other three methods.

The results of removing the contrast transform feature extraction part are depicted in column (b). These outputs offer valuable insights into the significance of feature extraction in the fusion process, as evidenced by the comparatively diminished effectiveness of detecting the infrared target in the fused image when compared to our proposed method. This outcome further substantiates the superior performance of our proposed method, highlighting its superiority over other techniques in terms of fusing infrared images.

(a) (b) (c) (d)

Figure 14. Ablation experiments on "00613D" in the MSRS dataset. The infrared target edge details have been marked in red boxes and the background details have been marked in green boxes.

The outputs of the removal of the spatial-domain logistic filter are examined in column (c). The fused image appears relatively darker compared to the others and exhibits a certain degree of color distortion. This observation suggests that the spatial-domain logistic filter plays a crucial role in achieving intensity balance and ensuring color accuracy in the fused image.

The outputs in column (d) demonstrate the results of removing both the feature extraction component of the contrast transform and the spatial-domain logistic filter. In this case, the fused image appears significantly darker and exhibits a more pronounced degree of color distortion, thereby emphasizing the crucial role played by these elements in the fusion process.

Overall, these outputs from four sets of experiments serve to demonstrate the effectiveness of our proposed method and the importance of each component in the fusion process. The comparison between these outputs and our proposed method highlights the direct effectiveness of our approach, making it a promising technique for fusing infrared images.

In order to demonstrate the effectiveness of the weighted wavelet transform in the low-frequency part fusion, a third set of ablation experiments was conducted using a direct fusion strategy and a linear average fusion strategy for comparison. The specific subjective and objective output results are shown below.

In Figure 15, the infrared target is indicated by the red boxes, and it can be observed that the direct fusion algorithm yielded edge blurring in its result. The background information highlighted in green boxes demonstrates that the linear averaging algorithm and the direct fusion algorithm had information loss.

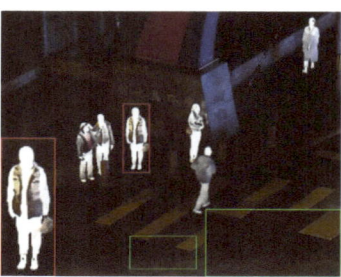 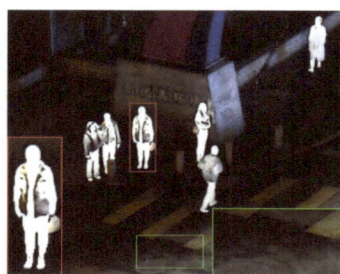 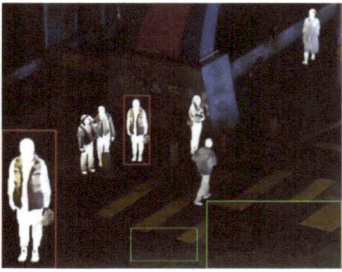

Ours | Direct fusion | Linear average fusion

Figure 15. Subjective results of ablation experiments. The infrared target edge details have been marked in red boxes and the background details have been marked in green boxes.

In order to further show the advantages of the weighted-wavelet fusion algorithm in information retention, we use ten evaluation indicators for objective tests: EN, AG, MI, VIF, Nabf, FMI, SF, SD, Qabf, and CC. The resulting output was as follows. Due to limited space, only two decimal places were retained. Moreover, the direct fusion is referred to as DF, and the linear average fusion is referred to as LF.

As illustrated in Table 5, the optimal values, namely, EN, AG, MI, Nabf, FMI, SF, and Qabf, significantly emphasize the efficacy of the wavelet-weighted fusion strategy for the low-frequency part fusion. This strategy not only effectively preserved the source image information but also generated a discernible fusion outcome, capturing minute visual details from the input image. Furthermore, the strategy enhanced the overall quality of the fused image by incorporating the wavelet transform, which inherently focused on the low-frequency components of the image. Consequently, the fused image exhibited a higher level of detail and a more accurate representation of the original source images, demonstrating the superiority of the wavelet-weighted fusion strategy over other fusion methods. To visualize the difference between the results of the various methods, we present a histogram in Figure 16.

Table 5. Objective evaluation metrics for the third set of ablation experiments on the LLVIP dataset.

Algo	EN	AG	MI	VIF	Nabf	FMI	SF	SD	Qabf	CC
DF	6.38	1.78	3.42	**1.16**	0.23	0.91	8.26	7.84	0.58	**0.82**
LF	6.37	1.95	4.03	1.03	**0.08**	0.93	8.12	**8.86**	0.56	0.80
Ours	**6.39**	**2.12**	**4.33**	1.06	**0.08**	**0.94**	**8.39**	7.87	**0.64**	0.81

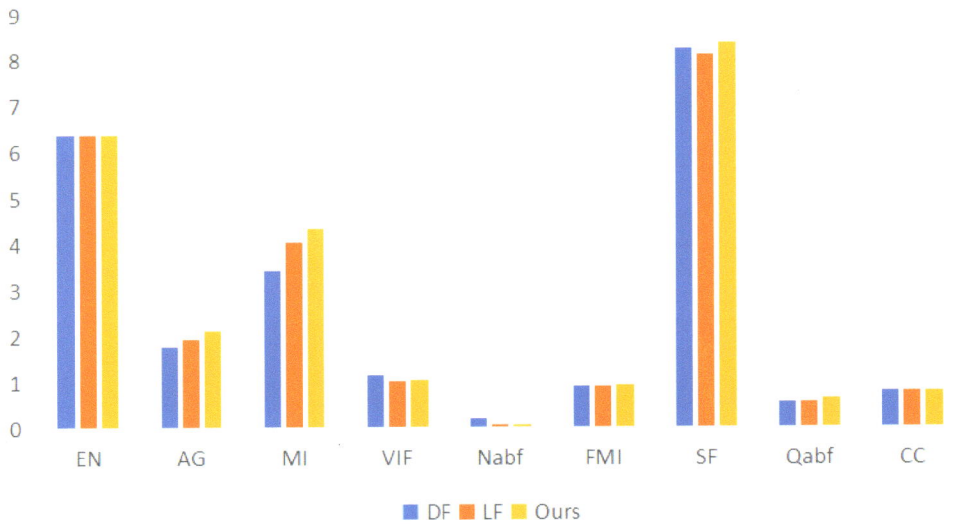

Figure 16. Histogram of the third set of objective evaluation indicators from the ablation experiment.

4.5. Real-Time Validation Experiment

In order to verify that DDCTFuse had good visual output effect and real-time performance, we compared DDCTFuse with some current advanced image fusion algorithms (RFN-Net, DenseFuse, NestFuse, and LatLRR) on the LLVIP and MSRS datasets, mainly in terms of CPU or GPU usage, RAM usage, and algorithm response time. The specific experimental results are shown in Table 6. It should be noted that the above-mentioned deep learning methods all completed the model training and reasoning process in the same virtual environment (python-3.8.10; torch-2.2.0+cu121; torchvision-0.17.0). The specific experimental results are shown in Table 6.

Table 6. Performance comparison of algorithms on different datasets.

Dataset	Algorithm	CPU	GPU	RAM	Time (s)
MSRS	LatLRR	65%	—	51 MB	0.331
	RFN-Net	—	98%	3.5 GB	0.836
	DenseFuse	—	100%	0.7 GB	0.451
	NestFuse	—	96%	1.3 GB	0.764
	Ours	14%	—	32 MB	0.051
LLVIP	LatLRR	47%	—	67 MB	0.384
	RFN-Net	—	100%	3.7 GB	0.913
	DenseFuse	—	99%	0.8 GB	0.414
	NestFuse	—	99%	1.5 GB	0.815
	Ours	17%	—	34 MB	0.049

From the above experimental results, it can be concluded that the DDCTFuse method proposed in this paper achieved good performance in real time. When DDCTFuse was running, the CPU usage and the RAM usage was minimal, the algorithm response time was short, and the output result was fast. The above advantages prove that DDCTFuse is very advantageous in actual application scenarios.

5. Discussion

Combining the subjective and objective evaluations in the previous section and the results of the ablation experiments, we proved that the new contrast transform function we designed had excellent performance in the IR target extraction task, which can be reflected by the value of VIF and FMI. The spatial-domain logic optimized filter proposed in this paper also played an equally important role, as can be seen by the larger MI and AG.

Looking ahead, we plan to further refine the nonlinear element function for the contrast transform and extend its application to other computer vision tasks such as edge detection and image segmentation. We anticipate that this expansion will not only enhance the performance of these tasks but also broaden the scope of our research. Additionally, we are considering the integration of the logical filter into deep learning methods for color correction at various feature levels. This interdisciplinary approach holds the potential to make substantial advancements in the field of computer vision and open new avenues for research and development.

6. Conclusions

In this paper, we proposed an infrared and visible image fusion strategy (DDCTFuse) based on an adaptive high-pass filter, a logical filter, and contrast transform feature extraction, which enhanced the infrared target and transformed problems that are difficult to deal with in the frequency domain, such as ringing artifacts, into the spatial-domain by using a spatial-domain logical filter for processing. The experimental results showed that DDCTFuse achieved superior visual effects and evaluation results with strong generalization ability and good robustness when compared with nine fusion algorithms on four public datasets.

Author Contributions: Conceptualization, X.M. and T.L. (Tianqi Li); methodology, X.M. and T.L. (Tianqi Li); software, T.L. (Tianqi Li); validation, S.H., T.L. (Tianqi Li) and J.D.; formal analysis, X.M., S.H. and J.D.; investigation, T.L. (Tianqi Li); resources, X.M. and T.L. (Tianqi Li); data curation, X.M. and T.L. (Tianqi Li); writing—original draft preparation, T.L.; writing—review and editing, T.L. (Tianqi Li), X.M., S.H. and J.D.; visualization, T.L. (Tianqi Li), T.L. (Tong Li) and J.D.; supervision,

R.W., C.C., T.Q., G.L. and J.L. (Tong Li); project administration, X.M. and S.H.; funding acquisition, S.H., X.M. and J.D. All authors have read and agreed to the published version of the manuscript.

Funding: This work was funded by the China Postdoctoral Science Foundation under Grant 2020M683522, and the Natural Science Basic Research Program of Shanxi under Grant 2024JC-YBMS-490.

Institutional Review Board Statement: Not applicable.

Informed Consent Statement: Not applicable.

Data Availability Statement: The data presented in this study are available on request from the corresponding author due to privacy.

Conflicts of Interest: The authors declare no conflicts of interest.

References

1. Luo, Y.; Wang, X.; Wu, Y.; Shu, C. Infrared and Visible Image Homography Estimation Using Multiscale Generative Adversarial Network. *Electronics* **2023**, *12*, 788. [CrossRef]
2. Ji, J.; Zhang, Y.; Lin, Z.; Li, Y.; Wang, C.; Hu, Y.; Huang, F.; Yao, J. Fusion of Infrared and Visible Images Based on Optimized Low-Rank Matrix Factorization with Guided Filtering. *Electronics* **2022**, *11*, 2003. [CrossRef]
3. Tu, Z.; Li, Z.; Li, C.; Lang, Y.; Tang, J. Multi-interactive dual-decoder for RGB-thermal salient object detection. *IEEE Trans. Image Process.* **2021**, *30*, 5678–5691. [CrossRef]
4. Nagarani, N.; Venkatakrishnan, P.; Balaji, N. Unmanned Aerial vehicle's runway landing system with efficient target detection by using morphological fusion for military surveillance system. *Comput. Commun.* **2020**, *151*, 463–472. [CrossRef]
5. Vidas, S.; Moghadam, P.; Bosse, M. 3D thermal mapping of building interiors using an RGB-D and thermal camera. In Proceedings of the 2013 IEEE International Conference on Robotics and Automation, Karlsruhe, Germany, 6–10 May 2013; IEEE: Piscataway, NJ, USA, 2013; pp. 2311–2318.
6. Dinh, P.H. Combining gabor energy with equilibrium optimizer algorithm for multi-modality medical image fusion. *Biomed. Signal Process. Control* **2021**, *68*, 102696. [CrossRef]
7. Ma, J.; Ma, Y.; Li, C. Infrared and visible image fusion methods and applications: A survey. *Inf. Fusion* **2019**, *45*, 153–178. [CrossRef]
8. Zhao, L.; Zhang, Y.; Dong, L.; Zheng, F. Infrared and visible image fusion algorithm based on spatial-domain and image features. *PLoS ONE* **2022**, *17*, e0278055. [CrossRef]
9. Huang, Y.; Bi, D.; Wu, D. Infrared and visible image fusion based on different constraints in the non-subsampled shearlet transform domain. *Sensors* **2018**, *18*, 1169. [CrossRef]
10. Ma, J.; Chen, C.; Li, C.; Huang, J. Infrared and visible image fusion via gradient transfer and total variation minimization. *Inf. Fusion* **2016**, *31*, 100–109. [CrossRef]
11. Saeedi, J.; Faez, K. Infrared and visible image fusion using fuzzy logic and population-based optimization. *Appl. Soft Comput.* **2012**, *12*, 1041–1054. [CrossRef]
12. Li, H.; Liu, L.; Huang, W.; Yue, C. An improved fusion algorithm for infrared and visible images based on multi-scale transform. *Infrared Phys. Technol.* **2016**, *74*, 28–37. [CrossRef]
13. Yu, S.; Chen, X. Infrared and visible image fusion based on a latent low-rank representation nested with multiscale geometric transform. *IEEE Access* **2020**, *8*, 110214–110226. [CrossRef]
14. Chen, J.; Li, X.; Luo, L.; Mei, X.; Ma, J. Infrared and visible image fusion based on target-enhanced multiscale transform decomposition. *Inf. Sci.* **2020**, *508*, 64–78. [CrossRef]
15. Pu, T.; Ni, G. Contrast-based image fusion using the discrete wavelet transform. *Opt. Eng.* **2000**, *39*, 2075–2082. [CrossRef]
16. Li, C.; Lei, L.; Zhang, X. Infrared and Visible Image Fusion Based on Morphological Enhancement of Dual-Tree Complex Wavelet. In Proceedings of the Advances in Natural Computation, Fuzzy Systems and Knowledge Discovery: Volume 2, Xi'an, China, 1–3 August 2020; Springer: Cham, Switzerland, 2020; pp. 743–752.
17. Li, H.; Qiu, H.; Yu, Z.; Zhang, Y. Infrared and visible image fusion scheme based on NSCT and low-level visual features. *Infrared Phys. Technol.* **2016**, *76*, 174–184. [CrossRef]
18. Tan, W.; Zhang, J.; Xiang, P.; Zhou, H.; Thitøn, W. Infrared and visible image fusion via NSST and PCNN in multiscale morphological gradient domain. In Proceedings of the Optics, Photonics and Digital Technologies for Imaging Applications VI, Online, 6–10 April 2020; SPIE: Paris, France, 2020; Volume 11353, pp. 297–303.
19. Johnson, J.L.; Padgett, M.L. PCNN models and applications. *IEEE Trans. Neural Netw.* **1999**, *10*, 480–498. [CrossRef]
20. Junwu, L.; Li, B.; Jiang, Y. An infrared and visible image fusion algorithm based on LSWT-NSST. *IEEE Access* **2020**, *8*, 179857–179880. [CrossRef]
21. Suryanarayana, G.; Varadarajan, V.; Pillutla, S.R.; Nagajyothi, G.; Kotapati, G. Multiple Degradation Skilled Network for Infrared and Visible Image Fusion Based on Multi-Resolution SVD Updation. *Mathematics* **2022**, *10*, 3389. [CrossRef]
22. He, C.; Liu, Q.; Li, H.; Wang, H. Multimodal medical image fusion based on IHS and PCA. *Procedia Eng.* **2010**, *7*, 280–285. [CrossRef]

23. Wang, J.; Lai, S.; Li, M. Improved image fusion method based on NSCT and accelerated NMF. *Sensors* **2012**, *12*, 5872–5887. [CrossRef]
24. Wang, J.; Shi, D.; Cheng, D.; Zhang, Y.; Gao, J. LRSR: Low-rank-sparse representation for subspace clustering. *Neurocomputing* **2016**, *214*, 1026–1037. [CrossRef]
25. Abdolali, M.; Rahmati, M. Multiscale decomposition in low-rank approximation. *IEEE Signal Process. Lett.* **2017**, *24*, 1015–1019. [CrossRef]
26. Chua, L.O.; Roska, T. The CNN paradigm. *IEEE Trans. Circuits Syst. I Fundam. Theory Appl.* **1993**, *40*, 147–156. [CrossRef]
27. Bavirisetti, D.P.; Dhuli, R. Fusion of infrared and visible sensor images based on anisotropic diffusion and Karhunen-Loeve transform. *IEEE Sensors J.* **2015**, *16*, 203–209. [CrossRef]
28. Li, H.; Wu, X.J. Infrared and visible image fusion using latent low-rank representation. *arXiv* **2018**, arXiv:1804.08992.
29. Li, H.; Wu, X.J. DenseFuse: A fusion approach to infrared and visible images. *IEEE Trans. Image Process.* **2018**, *28*, 2614–2623. [CrossRef]
30. Zhang, Y.; Liu, Y.; Sun, P.; Yan, H.; Zhao, X.; Zhang, L. IFCNN: A general image fusion framework based on convolutional neural network. *Inf. Fusion* **2020**, *54*, 99–118. [CrossRef]
31. Tang, L.; Xiang, X.; Zhang, H.; Gong, M.; Ma, J. DIVFusion: Darkness-free infrared and visible image fusion. *Inf. Fusion* **2023**, *91*, 477–493. [CrossRef]
32. Li, H.; Wu, X.J.; Durrani, T. NestFuse: An infrared and visible image fusion architecture based on nest connection and spatial/channel attention models. *IEEE Trans. Instrum. Meas.* **2020**, *69*, 9645–9656. [CrossRef]
33. Li, H.; Wu, X.J.; Kittler, J. RFN-Nest: An end-to-end residual fusion network for infrared and visible images. *Inf. Fusion* **2021**, *73*, 72–86. [CrossRef]
34. Tang, L.; Yuan, J.; Ma, J. Image fusion in the loop of high-level vision tasks: A semantic-aware real-time infrared and visible image fusion network. *Inf. Fusion* **2022**, *82*, 28–42. [CrossRef]
35. Jia, X.; Zhu, C.; Li, M.; Tang, W.; Zhou, W. LLVIP: A visible-infrared paired dataset for low-light vision. In Proceedings of the IEEE/CVF International Conference on Computer Vision, Montreal, BC, Canada, 11–17 October 2021; pp. 3496–3504.
36. Ma, J.; Tang, L.; Fan, F.; Huang, J.; Mei, X.; Ma, Y. SwinFusion: Cross-domain Long-range Learning for General Image Fusion via Swin Transformer. *IEEE/CAA J. Autom. Sin.* **2022**, *9*, 1200–1217. [CrossRef]
37. Xu, H.; Ma, J.; Jiang, J.; Guo, X.; Ling, H. U2Fusion: A Unified Unsupervised Image Fusion Network. *IEEE Trans. Pattern Anal. Mach. Intell.* **2020**, *44*, 502–518. [CrossRef]

Disclaimer/Publisher's Note: The statements, opinions and data contained in all publications are solely those of the individual author(s) and contributor(s) and not of MDPI and/or the editor(s). MDPI and/or the editor(s) disclaim responsibility for any injury to people or property resulting from any ideas, methods, instructions or products referred to in the content.

Article

Research on Obstacle-Avoidance Trajectory Planning for Drill and Anchor Materials Handling by a Mechanical Arm on a Coal Mine Drilling and Anchoring Robot

Siya Sun [1,2], Sirui Mao [3], Xusheng Xue [1,2], Chuanwei Wang [1,2], Hongwei Ma [1,2,*], Yifeng Guo [1,2], Haining Yuan [3] and Hao Su [1,2]

[1] College of Mechanical Engineering, Xi'an University of Science and Technology, Xi'an 710054, China; sunsiya412@xust.edu.cn (S.S.); xuexsh@xust.edu.cn (X.X.); wangchuanwei228@xust.edu.cn (C.W.); guoyifeng000721@163.com (Y.G.); suh@stu.xust.edu.cn (H.S.)

[2] Shaanxi Key Laboratory of Mine Electromechanical Equipment Intelligent Detection and Control, Xi'an 710054, China

[3] College of Electrical and Control Engineering, Xi'an University of Science and Technology, Xi'an 710054, China; 24206204126@stu.xust.edu.cn (S.M.); 21406050426@stu.xust.edu.cn (H.Y.)

* Correspondence: mahw@xust.edu.cn

Abstract: At present, China's coal mine permanent tunneling support commonly uses mechanized drilling and anchoring equipment; there are low support efficiency, labor intensity, and other issues. In order to further improve the support efficiency and liberate productivity, this paper further researches the trajectory planning of the drilling and anchoring materials of the robotic arm for the drilling machine "grasping–carrying–loading–unloading" on the basis of the drilling and anchoring robotic system designed by the team in the previous stage. Firstly, the kinematic model of the robotic arm with material was established by improving the D-H parameter method. Then, the working space of the robotic arm with the material was analyzed using the Monte Carlo method. The singular bit-shaped region of the robotic arm was restricted, and obstacles were removed from the working space. The inverse kinematics was utilized to solve the feasible domain of the robotic arm with material. Secondly, in order to avoid blind searching, the guidance of the Bi-RRT algorithm was improved by adding the target guidance factor, and the two-way tree connection strategy for determining the feasible domain was combined with the Bi-RRT algorithm's feasible domain judgment bi-directional tree connection strategy to improve the convergence speed of the Bi-RRT algorithm. Then, in order to adapt to the dynamic environment and avoid the global planning algorithm from falling into the local minima, on the basis of the above planning methods, an improved Bi-RRT trajectory planning algorithm incorporating the artificial potential field was proposed, which takes the planned paths as the guiding potential field of the artificial potential field to make full use of the global information and avoid falling into the local minimization. Finally, a simulation environment was built in a ROS environment to compare and analyze the planning effect of different algorithms. The simulation results showed that the improved Bi-RRT trajectory planning algorithm incorporating the artificial potential field improved the optimization speed by 69.8% and shortened the trajectory length by 46.6% compared with the traditional RRT algorithm.

Keywords: anchor drilling robot; material handling manipulator arm; dynamic obstacle avoidance; trajectory planning

Citation: Sun, S.; Mao, S.; Xue, X.; Wang, C.; Ma, H.; Guo, Y.; Yuan, H.; Su, H. Research on Obstacle-Avoidance Trajectory Planning for Drill and Anchor Materials Handling by a Mechanical Arm on a Coal Mine Drilling and Anchoring Robot. *Sensors* **2024**, *24*, 6866. https://doi.org/10.3390/s24216866

Academic Editor: Sergio Toral Marín

Received: 11 September 2024
Revised: 18 October 2024
Accepted: 24 October 2024
Published: 25 October 2024

Copyright: © 2024 by the authors. Licensee MDPI, Basel, Switzerland. This article is an open access article distributed under the terms and conditions of the Creative Commons Attribution (CC BY) license (https://creativecommons.org/licenses/by/4.0/).

1. Introduction

According to statistics, the support time in a tunneling cycle operation in a coal mine roadway accounts for about 2/3 of the total time. Therefore, improving the intelligence of the permanent support equipment in the roadway is an important way to improve the efficiency of tunneling. Since 2015, the state ministries and commissions have successively issued some technical guidelines to enhance the level of coal mine development and improve the level of safety production, which is of great significance for the transformation of the permanent support equipment of the underground roadway of coal mines to intelligence [1–3]. Automatic loading and unloading drilling and anchoring material technology is one of the key technologies in the research of intelligent drilling and anchoring platforms [4]. The common coal mine underground intelligent drilling anchor equipment takes a load-sensitive electro-hydraulic proportional multi-way valve as the main core, PLC as the core control unit, remote control as the operation terminal, and displacement sensor, pressure sensor, and so on, as the detecting and feedback elements [5–8]. A few drilling anchor platforms have multiple automatic loading and unloading drilling rod mechanisms, and their intelligent cooperative control part is still in the research stage. The existing equipment has problems such as complex structure, many detection elements, low positioning accuracy, large repetitive positioning errors, and a single loading and unloading material, which has a large impact on the reliability of support operations [9–11]. In this paper, on the basis of intelligent drilling and an anchoring robot structure with a multi-degree-of-freedom robotic arm proposed by the team in the previous stage, we study a method of obstacle avoidance trajectory planning for a robotic arm with material in a dynamic environment.

For the problem of robotic arm obstacle avoidance trajectory planning, Chen et al. [12] combined non-obstacle spatial sampling with the artificial potential field method to generate heuristic sampling points, which enabled the RRT trajectory planning algorithm to obtain higher efficiency and better trajectories. Xiao et al. [13] carried out video detection by a flexible manipulator along a continuous path in a narrow 3D space and generated the trajectory under a variety of constraints, such as the terminal camera attitude, drive space, and obstacles, to generate trajectories. Zhang et al. [14] proposed a path construction strategy to remove redundant nodes to improve the smoothness and reduce the memory storage of the algorithm based on an artificial potential field method and a two-way fast exploratory random tree path planning algorithm for solving the problem of high randomness and low search efficiency. Wang et al. [15] proposed a fusion algorithm of improved artificial potential field method (APF) and rapid extended random tree method (RRT) to address the problems of low success rate and low efficiency in obstacle avoidance path planning for 6R robotic arms in complex environments. The improved APF-RRT algorithm can adapt to complex environments and effectively solves the problems of unreachable APF targets and local minima. Chen et al. [16] proposed a hybrid artificial potential field and ant colony optimization based on raster maps for solving the path planning problem of robots in known environments. Chen et al. [17] proposed a Biased-RRT correction algorithm based on sampling rule goal-oriented design and parent node re-selection to solve the path planning problem of a six-degree-of-freedom robotic arm in a complex environment. This algorithm can effectively shorten the path planning time, reduce the path length, and better accomplish the expected path planning task of a six-degree-of-freedom robotic arm in a complex environment. Zhang et al. [18] designed a Bi-RRT algorithm guided by the artificial potential field method for surface unmanned boats, and the experiment proved that the algorithm can greatly reduce the path length and the number of nodes and improve the path smoothness while improving the search efficiency.

In summary, most of the existing robotic arm obstacle avoidance trajectory optimization methods consider the end material as a mass point, and there is little research on obstacle avoidance optimization for grasping rod materials with a large aspect ratio. In the coal mine tunnel drilling and anchoring support process, the drilling and anchoring materials are mainly long and thin rods, and it is difficult to plan the trajectory of the robotic arm with the material in the restricted space-time.

2. Design of the Anchoring Robot System

By analyzing the process of parallel cyclic operation of coal mine tunneling and permanent support, the team designed a drilling and anchoring robot system with a collaborative robotic arm in the early stage. The system consists of an anchor drilling platform, an automatic drilling machine, a multi-degree-of-freedom collaborative robotic arm, an anchor drilling material library, and an electronic control box. The structure is shown in Figure 1. Among them, the multi-degree-of-freedom collaborative robotic arm is based on the traditional six-axis robotic arm with the addition of a horizontal moving slide and a lifting platform, which is designed to be able to meet the workspace requirements of the anchor drilling task. The structure is shown in Figure 2.

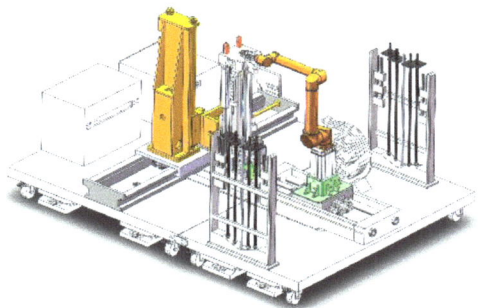

Figure 1. Structure of anchor drilling robot system.

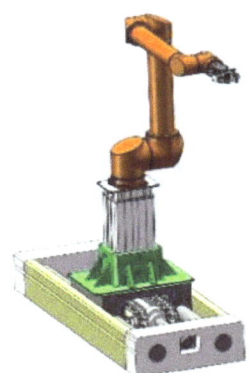

Figure 2. Overall structure of anchor drilling robot arm.

3. Kinematic Modeling and Feasible Domain Analysis of Material-Carrying Robotic Arm

The coal mine roadway drilling and anchoring support process is relatively cumbersome. To form the machine to complete the drilling and anchoring operations, we mainly need to solve the problem of loading and unloading of drilling and anchoring materials. As the drilling anchor materials are mostly long and thin rods, the mechanical arm clamps them in the drilling and anchoring platform delivery process. In order to reduce the probability of collision, the drilling anchor materials are required to move perpendicular to the platform to minimize its duty cycle. Therefore, the robotic arm with the material described in this paper refers to the robotic arm that clamps the drill-anchor material at the end, and special constraints are made on the attitude of this robotic arm in the process of delivering the material; the structure is shown in Figure 3.

Figure 3. Schematic diagram of a mechanical arm with material.

3.1. Kinematics Modeling of Material-Carrying Robotic Arm

In this study, the improved D-H method is used to model and analyze the connecting rod and joints of the robotic arm with material, and the rod-shaped material is regarded as the end connecting rod of the robotic arm with material. The coordinate system of the multi-degree-of-freedom collaborative robotic arm and the connecting rod of the robotic arm with the material are established, respectively, as shown in Figures 4 and 5, which are the three-dimensional coordinates of different joint points, respectively.

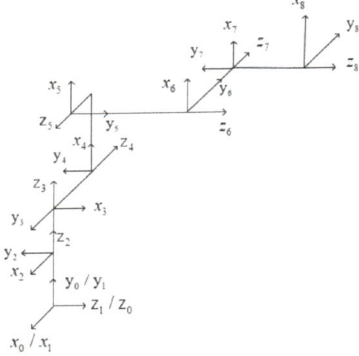

Figure 4. Coordinate diagram of connecting rod of the cooperative manipulator.

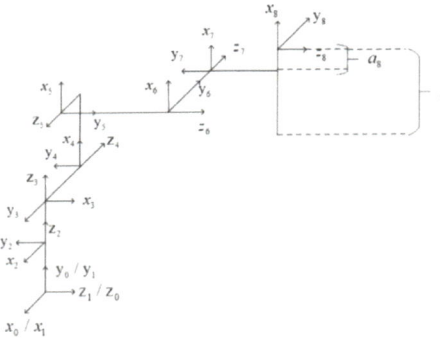

Figure 5. Drawing of connecting rod coordinate system of conveyor arm.

Based on the arm as well as the actual dimensions of the drilled anchor material, the DH parameter list of the banded robotic arm can be obtained, as shown in Table 1. a_{i-1} is the length of the linkage, α_{i-1} is the angle of torsion of the linkage, θ_i is the joint variable of the rotating joints, and d_i is the offset of the linkage.

Table 1. D-H parameters of drilling anchor robotic arm.

i	α_{i-1}	a_{i-1}	d_i	θ_i
1	0	0	d_1	0
2	90	0	d_2	0
3	0	0	10	θ_3
4	90	10	0	θ_4
5	−90	60	10	θ_5
6	90	0	60	θ_6
7	90	0	10	θ_7
8	−90	a_8	10	θ_8

The transformation relationship between the neighboring joints of the banding robot arm can be represented by the transformation matrix T. The transformation relationship between the coordinate system $\{i\}$ and coordinate system $\{i-1\}$ can be represented by the transformation matrix $^{i-1}_{i}T$:

$$^{i-1}_{i}T = \begin{bmatrix} c\theta_i & -s\theta_i & 0 & a_{i-1} \\ s\theta_i c\alpha_{i-1} & c\theta_i c\alpha_{i-1} & -s\alpha_{i-1} & -d_i s\alpha_{i-1} \\ s\theta_i s\alpha_{i-1} & c\theta_i s\alpha_{i-1} & c\alpha_{i-1} & d_i c\alpha_{i-1} \\ 0 & 0 & 0 & 1 \end{bmatrix} \quad (1)$$

According to the D-H parameter table of the robotic arm with material, the transformation matrix of the end rod material with respect to the base coordinate system is as follows:

$$\theta_5 = -\theta_7 \quad (2)$$

$$\theta_6 = -\theta_8 \quad (3)$$

After obtaining the displacement and inclination data of the robotic arm and substituting them into the transformation matrix, the spatial attitude of the robotic arm, as well as the drilling and anchoring material, can be obtained.

3.2. Collision Detection of Material-Carrying Robotic Arm

The motion of the robotic arm must be taken into account in the collision detection algorithm [19], where the kinematics are modeled in order to describe the state of motion of the robotic arm and the spatial position of the linkage. The accurate position of the robotic arm linkage obtained from the positive kinematics provides the basic data for collision detection. Subsequently, shape information is obtained by enveloping the robotic arm linkage and obstacles. Finally, it is calculated to determine whether a collision occurs or not. Because the components of the robotic arm are composed of irregular geometry, it is difficult to model accurately, and the calculation is too large. However, because the robotic arm connecting rod is similar to a cylinder, it can be used to reduce the difficulty of modeling and reduce the detection time. Regular, irregular obstacles similar to a sphere can be enveloped by a sphere, while rod-shaped obstacles can be enveloped by a number of spheres with a smaller radius. The specific modeling is shown in Figure 6.

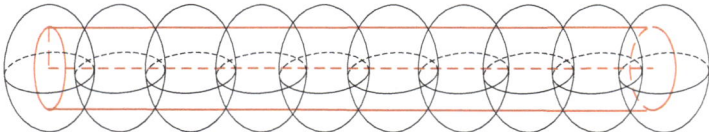

Figure 6. Simplified model of obstacles.

The collision detection between the robotic arm and the obstacle can be transformed into the collision detection between the enveloped sphere and the enveloped cylinder, based on which the radius of the cylinder can be superimposed into the radius of the sphere, which can be further transformed into the collision detection between the sphere and the line segments, as shown in Figure 7.

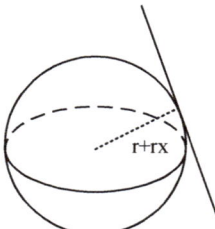

Figure 7. Collision of connecting rod with obstacle.

Let the cylinder that envelopes the linkage of the robotic arm have a center $A(x_1, y_1, z_1)$ at one end and a center $B(x_2, y_2, z_2)$ at the other end. Let them also have a radius of r_x and a sphere with a center C at the coordinates of the point $C(x_0, y_0, z_0)$ and a radius of r, and carry out the superposition after the sphere has a radius of $r_x + r$. Finally, let there be a point on the rod $D(x_d, y_d, z_d)$. According to the expression of the straight line in the space, the value of each coordinate of the point can be obtained:

$$\begin{cases} x_d = \lambda x_1 + (1-\lambda)x_2 \\ y_d = \lambda y_1 + (1-\lambda)y_2 \\ z_d = \lambda z_1 + (1-\lambda)z_2 \end{cases} \quad (4)$$

where λ is a variable between [0, 1], from which the distance from point D to the center of the sphere C is obtained:

$$d = \sqrt{(x_d - x_0)^2 + (y_d - y_0)^2 + (z_d - z_0)^2} \quad (5)$$

In the Formula (5), there is only one variable λ, which can build a function $f(\lambda) = d^2$ with a minimum value of $f(\lambda)$. Then, compared with the radius of the sphere, when $d^2_{min} > (r + r_x)^2$, the arm of the linkage did not collide with the obstacle, and vice versa, the collision occurred. Through the collision detection between different connecting rods and obstacles, we can finally determine whether the robotic arm has collided with the obstacle.

In addition to the collision between the robotic arm and the obstacle, the collision between the connecting rod and the connecting rod cannot be ignored due to its own structural properties. According to the structure of the six-axis robotic arm, the collision between the connecting rods generally occurs between connecting rod one and connecting rod three, connecting rod one and connecting rod four, and connecting rod two and connecting rod four. The collision between the connecting rods can be simplified as the collision between the cylinders, which is judged by calculating the distance between the axes of the two cylinders, as shown in Figure 8.

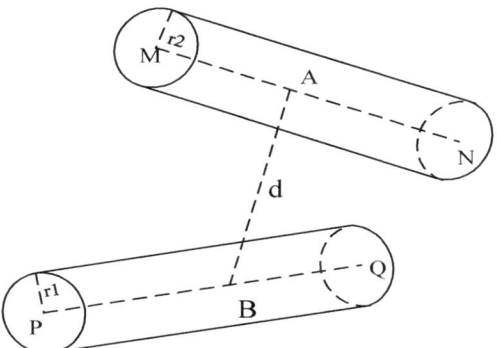

Figure 8. The collision between rods.

The cylinders of the envelope connecting rod one have centers M, N at each end, a radius r_2, and a point A on the axis. The cylinders of the envelope connecting rod 3 have centers Q, P, a radius r_1, and a point B on the axis, The distance between AB is d. The equation for the coordinates of points A and B can be obtained:

$$\begin{cases} A = M + \lambda(N - M) \\ B = P + \mu(Q - P) \end{cases} \tag{6}$$

where λ, μ are both variables between [0, 1], and the distance between AB can be calculated according to the coordinates of A, B points:

$$d = \sqrt{(x_A - x_B)^2 + (y_A - y_B)^2 + (z_A - z_B)^2} \tag{7}$$

The variables in Equation (7) are λ, μ. The binary function $f(\lambda, \mu) = d^2$ is constructed, and the $d^2{}_{min}$ distance between the two axes is obtained when $f(\lambda, \mu)$ takes the minimum value, while the minimum value of $f(\lambda, \mu)$ can be obtained by the Lagrange multiplier method:

$$\begin{cases} \frac{\partial f(\lambda, \mu)}{\partial \lambda} = 0 \\ \frac{\partial f(\lambda, \mu)}{\partial \mu} = 0 \end{cases} \tag{8}$$

Solve for different sets of values of λ, μ according to the conditions of Equation (8), and substitute into Equation (7) to find the final d_{min}. When $d_{min} > r_1 + r_2$ it can be assumed that there is no collision between connecting rod I and connecting rod III. Similarly, by calculating whether a collision occurs between connecting rod one and connecting rod four and between connecting rod two and connecting rod four, it can be determined whether a collision occurs between connecting rods of the robotic arm. When there is no collision between the robotic arm and the obstacle or between the robotic arm's connecting rods, it can be assumed that the robotic arm has not collided in the space.

3.3. Feasibility Domain Analysis of Robotic Arm with Material

In general, the workspace of a robotic arm is defined as all the locations that can be reached by the end device after it has completed all operations, i.e., the total amount of space covered by the end-effector. A random probability-based digitization method, the Monte Carlo algorithm [20], is chosen. In this paper, the range of locations that can be reached by the robotic arm with a material is called the task space.

When a robotic arm operates in the workspace, due to its own inherent nature, certain special positions or configurations will occur so that the robotic arm loses certain degrees of freedom or fails to move normally, which is also known as the robotic arm singularity avoidance problem [21]. Due to the kinematic inverse solution, singularities generally only exist under trajectory planning in Cartesian space, and planning under joint space generally does not have singularities because it does not involve inverting the speed of the robotic arm. Under the Cartesian coordinate system, different singularities can be categorized into boundary singularities and interior singularities according to different singular bit shapes, as shown in Figure 9.

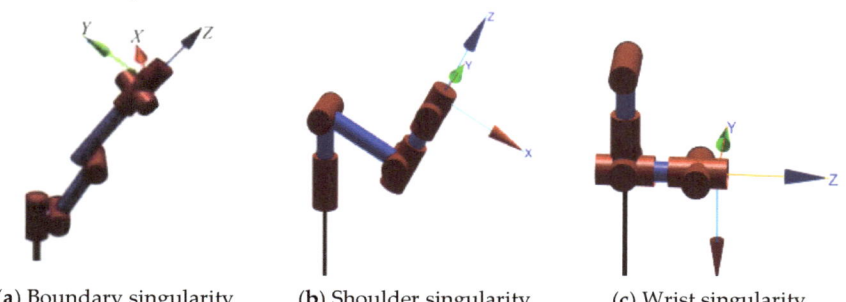

(a) Boundary singularity   (b) Shoulder singularity   (c) Wrist singularity

Figure 9. Different singular bit patterns.

In order to avoid the above situation and better solve the feasible domain of the robotic arm, the robotic arm can be restricted according to the singularity interval of the robotic arm, so that the robotic arm moves to avoid the joint angle in the range of this interval. As for the singular point in the operation of the robotic arm, it needs to be separated from the Jacobi matrix. The Jacobi determinant equal to zero is a sufficient condition for the singular point, and the singular region of the robotic arm can be solved by judging through this condition.

After restricting the robotic arm according to the singular interval, the obstacle interval can be removed on the basis of this to get the final position that meets the inherent properties of the robotic arm itself and the surrounding environment, i.e., the feasible domain of the robotic arm. After obtaining the feasible domain of the robotic arm, the planning algorithm can first determine whether the planned trajectory points are in the feasible domain to reduce the subsequent unnecessary judgment and accelerate the convergence of the algorithm. In the ROS simulation, the workspace of the robotic arm to keep the vertical attitude of the end rod is calculated, and the simulation results are shown in Figure 10. Among them, the yellow border is the workspace, the purple border is the task space, and the green border is the feasible domain.

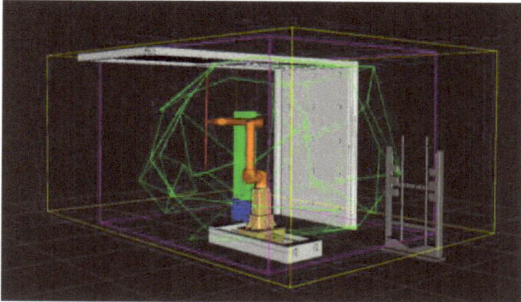

Figure 10. Feasible domain of the robotic arm.

4. Obstacle Avoidance Trajectory Planning Strategy for Robotic Arm with Material in Dynamic Environment

4.1. Bi-RRT Improvement Strategies

4.1.1. Connection Judgement

Using a fixed distance as a condition for judging whether a connection is made or not can lead to a degradation of the algorithm's performance under certain circumstances. Therefore, the algorithm in this study introduces a new connection judgment strategy. A new connection evaluation process is performed whenever a node is created, and the detailed steps are as follows: if both T_a and T_b are expanding a new node, the nodes generated by each are tried to be connected, and their possible routes are detected for collision. Once any obstacle is found blocking this route, the process of connecting is terminated, and the new node expansion is sought again. Only when no collision is encountered is it confirmed that the optimal solution has been found and the trajectory planning is completed, thus stopping the next expansion. The specific process is shown in Figure 11.

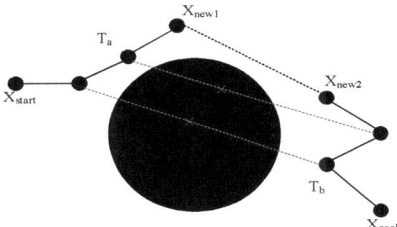

Figure 11. Connection Judgment Strategy.

As there will be a large number of calculations to perform collision detection, in order to reduce the amount of computation, set a suitable starting judgment distance value ds roughly for the starting point value of the target point distance of the general. This can only occur when the distance between the two new nodes is less than the value of the above connection operation.

4.1.2. Goal Oriented

Bi-RRT is the same as the RRT algorithm; although the two-way search strategy makes the algorithm converge much faster, the growth of nodes is still random and without goals, the problem of node redundancy still exists, and the growth of the two trees may occur in the opposite direction, which reduces the efficiency of the whole algorithm [22]. In this paper, for the algorithm under study, the goal-oriented strategy is invoked, so that T_a and T_b are expanded to the goal point and the starting point, respectively, which guides the two trees to meet and accelerates the algorithm's convergence. At the same time, in order to avoid the algorithm losing randomness, resulting in insufficient space for searching and falling into the local optimum, set a fixed guidance factor P, each time before generating a new node to randomly generate P_r for comparison with P. The specific guidance formula is shown in the following equation.

$$X_{new} = \begin{cases} X_{goal}/X_{start} & P_r > P \\ X_{rand} & P_r < P \end{cases} \quad (9)$$

When $P_r > P$, the newly generated nodes are made to point to the starting point or the goal point, depending on the tree. When $P_r < P$, generate new nodes by random sampling in the space. The specific process is shown in Figure 12.

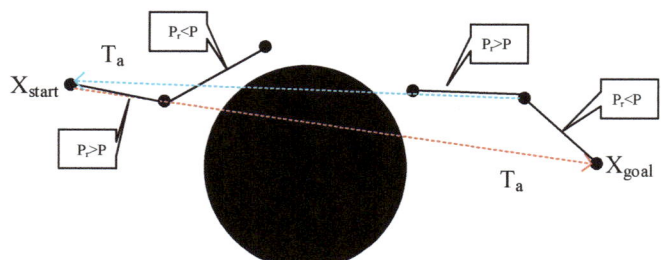

Figure 12. Goal Oriented Strategy.

4.1.3. Trajectory Simplification

The above improvement ideas can greatly reduce the speed of convergence of the algorithm, but even with the goal-oriented strategy, due to the algorithm's strong randomness, there are still a large number of redundant points and inflection points in the final trajectory, which is more obvious in the more complex obstacles. These redundant points and inflection points will make the robot arm take a redundant track and unnecessary steering during the actual movement, which will seriously cause irreversible loss of the robot arm.

In order to eliminate redundant points and superfluous inflection points, the planned trajectory needs to be simplified. The specific simplification process is shown in Figure 13.

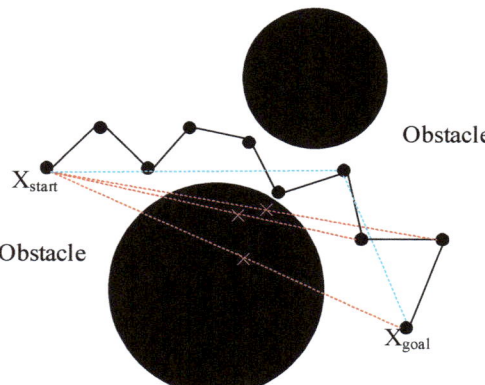

Figure 13. Trajectory simplification strategy.

For an already planned trajectory, a direct connection is made from the starting point X_{start} to the goal point X_{goal}, and if no collision occurs in the trajectory, this trajectory is taken as the final simplified trajectory. If a collision occurs, the trajectory is directly connected from X_{start} to the previous node of X_{goal}, and if a collision occurs in this trajectory it is continued from X_{start} to the previous node of this node. Instead, the node is used as the new parent node, and the above steps are repeated from that parent node until the parent node is directly connected to the goal point, representing the completion of the simplification of the trajectory.

4.2. Artificial Potential Field Improvement Strategies

4.2.1. Rewriting the Potential Field Function

A typical problem of the traditional artificial potential field algorithm is that the force field is unreasonable, i.e., the problem of too small gravitational force and too large repulsive force occurs in some special locations [23]. For this situation above, the function of gravitational force and repulsive force to optimize are rewritten. The rewritten gravitational field function and gravitational force function are shown in Equations (10) and (11), respectively.

$$U_{aat}(q) = \begin{cases} \frac{1}{2}K_{aat}\rho(q, q_{goal})^2, \rho(q, q_{goal}) \leq d \\ dK_{aat}\rho(q, q_{goal}) - \frac{1}{2}K_{aat}d^2, \rho(q, q_{goal}) > d \end{cases} \quad (10)$$

$$F_{aat}(q) = -\nabla U_{aat}(q) = \begin{cases} -K_{aat}\rho(q, q_{goal}), \rho(q, q_{goal}) \leq d \\ -dK_{aat}\frac{\partial \rho(q, q_{goal})}{\partial q}, \rho(q, q_{goal}) > d \end{cases} \quad (11)$$

where d is the distance factor, which sets a particular distance, and the rest of the parameters have the same meaning as before. According to this formula, when the position is farther away from the target point, the gravitational force received is reduced, and the magnitude of the reduction is distance-dependent.

4.2.2. Add Target Points

One of the notable problems with the artificial potential field algorithm is its inability to reach the goal point, which is mainly due to the fact that the repulsive and gravitational forces exerted during the motion continue to change, resulting in zero combined force at a particular location. This causes the algorithm to mistakenly believe that it has reached the goal point and, therefore, stops moving, which ultimately leads to the planning algorithm failing to successfully reach the goal point [24,25].

In order to avoid the artificial potential field algorithm suggesting the target point is not reachable, in the case of a combined force of zero, the method of adding random target points can be used. Through the setting of random target points to break the original force balance, so that the combined force is not zero, continue to move to the target point. Random target setting can be used in order to ensure that it can continue to move at the same time. However, it is necessary to consider if the gravitational force is too large to cause a collision with the obstacles. To increase the principle of the target point in the current position on the obstacle with the largest force to take the right symmetry, the distance needs to be greater than the obstacle of the repulsive force range. Increasing the target point will allow it to continue to exist until it is out of the repulsive force range of the obstacle, as shown in Figure 14.

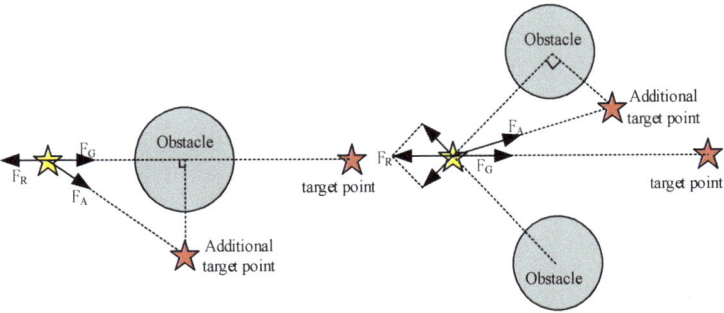

Figure 14. Add Target Points.

5. Obstacle Avoidance Trajectory Planning Algorithm Based on Improved Artificial Potential Field Fusion with Bi-RRT

Due to the existence of the local minimum problem, the trajectories planned by the improved artificial potential field algorithm are not optimal and have a large number of useless routes, while the improved Bi-RRT algorithm has fast planning speed and better global trajectories. In order to obtain a trajectory planning algorithm suitable for dynamic environments with better planning trajectories, the combination of the improved Bi-RRT algorithm and the improved artificial potential field algorithm is considered. The fusion algorithm is divided into two stages, namely static global trajectory planning and dynamic local trajectory planning. That is, the improved Bi-RRT algorithm is first utilized to carry out global trajectory planning in the static environment, and a globally better trajectory can be quickly obtained due to the short planning time and excellent planning route of the improved Bi-RRT algorithm. Subsequently, the improved artificial potential field algorithm is utilized to obtain dynamic obstacle information in real-time, and the global trajectory obtained in the static global trajectory planning stage is used to establish a guided potential field to avoid obstacles and converge to the global trajectory in real-time. The global planning trajectory of the bootstrap potential field is as follows:

$$U_{att}(q_{all}) = K_{rrt}\rho(q, q_{all})^2 \qquad (12)$$

where $\rho(q, q_{all})^2$ is the square of the shortest distance between the current position and the global trajectory, and K_{rrt} is the gravitational gain coefficient of the guiding potential field.

The gravitational potential field of the fusion algorithm consists of the gravitational potential field of the target point and the gravitational potential field of the global trajectory:

$$U_{att} = U_{att}(q_{goal}) + U_{att}(q_{all}) \qquad (13)$$

The specific workflow is as follows:

(1) Set the parameters of the banding robotic arm and obstacles, establish the initial position X_{start} and the target position X_{goal}, and initialize the growth trees T1 and T2. Solve the workspace of the banding robotic arm according to the positive kinematics, regard the singular interval of the banding robotic arm as the obstacle interval, and calculate the feasible domain N of the banding robotic arm by combining the obstacle information;

(2) Start alternating growth from the T1 tree or T2 tree as the current growing tree, generate a random factor P_r, and calculate the minimum distance between the newest node of the current tree and the nearest obstacle d. Determine the direction of expansion according to P_r, generate a new node X_{new}, and check whether it is within the feasible domain N;

(3) Envelope the robotic arm with material and obstacles and check the collision situation. If there is a collision, then go back to step 2. If there is no collision, then calculate the distance between the latest node of the two trees. If it is less than the value of the starting judgment distance ds, try to connect the two trees. If no collision occurs in the connected trajectory, then the planning trajectory is found, and the trajectory simplification should be carried out;

(4) According to the simplified trajectory, the obstacle information can be obtained in real-time and the gravitational field and repulsive field function can be constructed. Judge whether the combined force is zero at that time. If it is zero, then add a target point and enter step (5). Instead, advance to the next point and enter step (6);

(5) Advance to the next point and determine whether the current position escapes from the repulsive force range of the obstacle referenced by the incremental target point. If it does, remove the incremental target point and go back to step (4), and vice versa, repeat step (5);

(6) Determine whether the target point is reached. If so, end the algorithm and find the complete planning trajectory; otherwise, return to step (4) to continue the execution.

The flow chart of the obstacle avoidance trajectory planning algorithm based on the fusion of improved artificial potential field and Bi-RRT is shown in Figure 15.

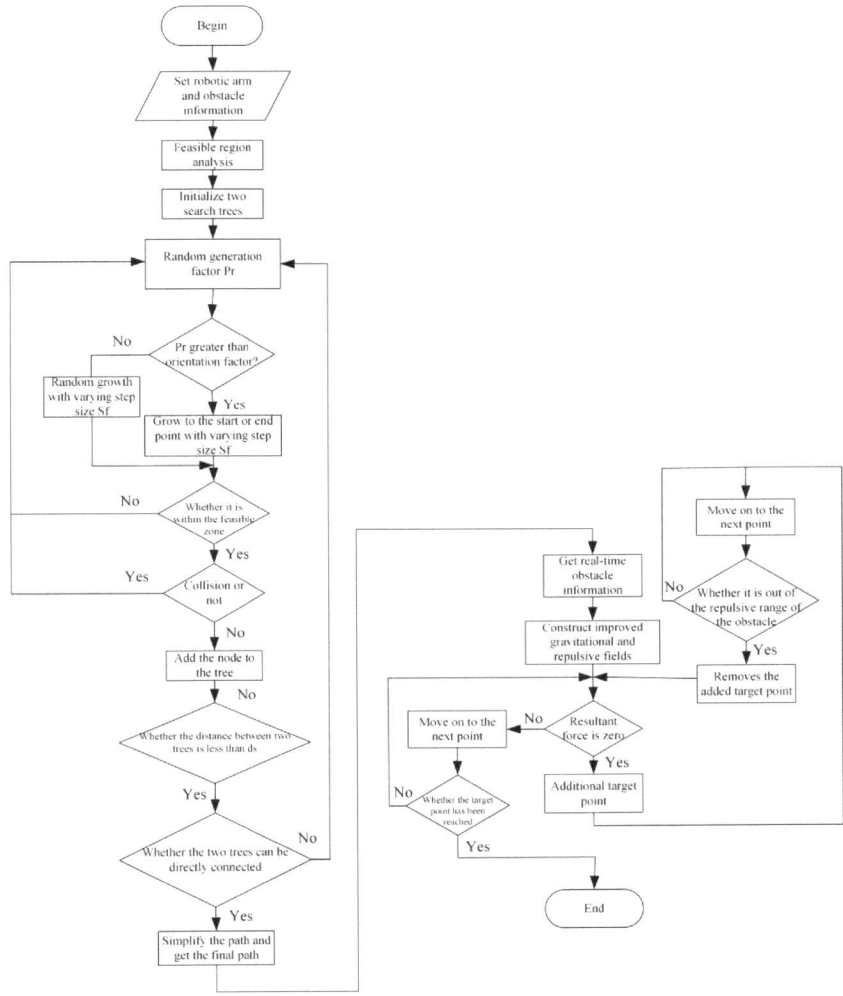

Figure 15. Based on the flow chart of the fusion algorithm of improved artificial potential field and Bi-RRT.

6. Experimental Validation

6.1. Theoretical Validation

When designing the artificial potential field, the setting and adjustment of parameters are very important for achieving good obstacle avoidance. The parameter settings need to be reasonably selected according to the specific application scene.

In order to compare and analyze the improvement of the fusion algorithm, the fusion algorithm and the improved artificial potential field algorithm are compared in MATLAB. The parameters of the algorithm are shown in Table 2.

A circular obstacle with the center of the circle [4, 4], [4, 6], [6, 4], a radius of 0.5 m is set in the environment. The final simulation result is shown in Figure 16.

Table 2. Algorithm parameters setting.

Parameter	Value
Gravitational coefficient K_a	28
Repulsion coefficient K_r	15
Repulsive force influence range d_0/m	2
Step size/m	0.1
Maximum number of iterations	200

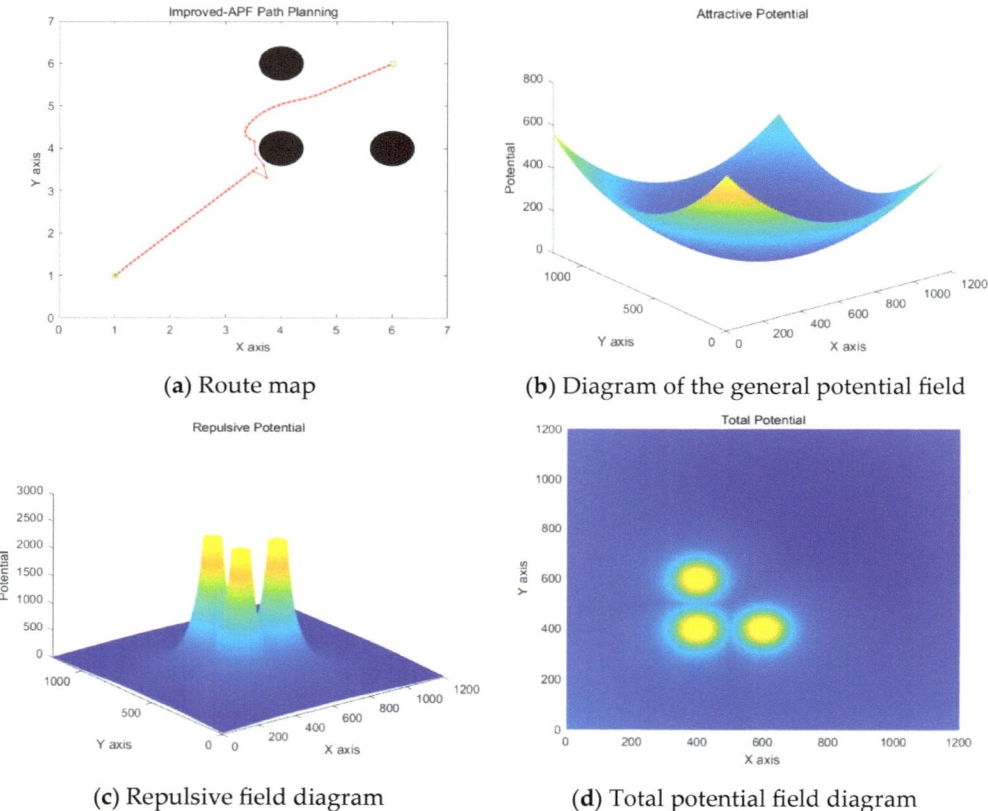

Figure 16. Improved artificial potential field simulation diagram.

According to the improved artificial potential field roadmap in Figure 16a, it can be found that the algorithm is not within the obstacle repulsive range at the beginning and is only affected by the gravitational force and the direction of the combined force points to the endpoint; the beginning trajectory is close to a straight line. When entering the repulsive range of the obstacle, there is a local minimum, reciprocating movement in a certain area, resulting in trajectory redundancy. In the planning effect diagram of different algorithms in Figure 17, the RRT algorithm performs the worst, with a large number of transitions during the growth of its nodes, resulting in a long expansion time. The RRT* algorithm has progressive optimization characteristics, which makes the trajectory length relatively shorter and smoother overall, but it also means that its convergence time will increase. The Bi-RRT trajectory is similar to the RRT, and there are more inflection points in the nodes, but the algorithm convergence time is faster. The fusion algorithm, on the other hand, outperforms the comparison algorithm in terms of trajectory length and trajectory quality.

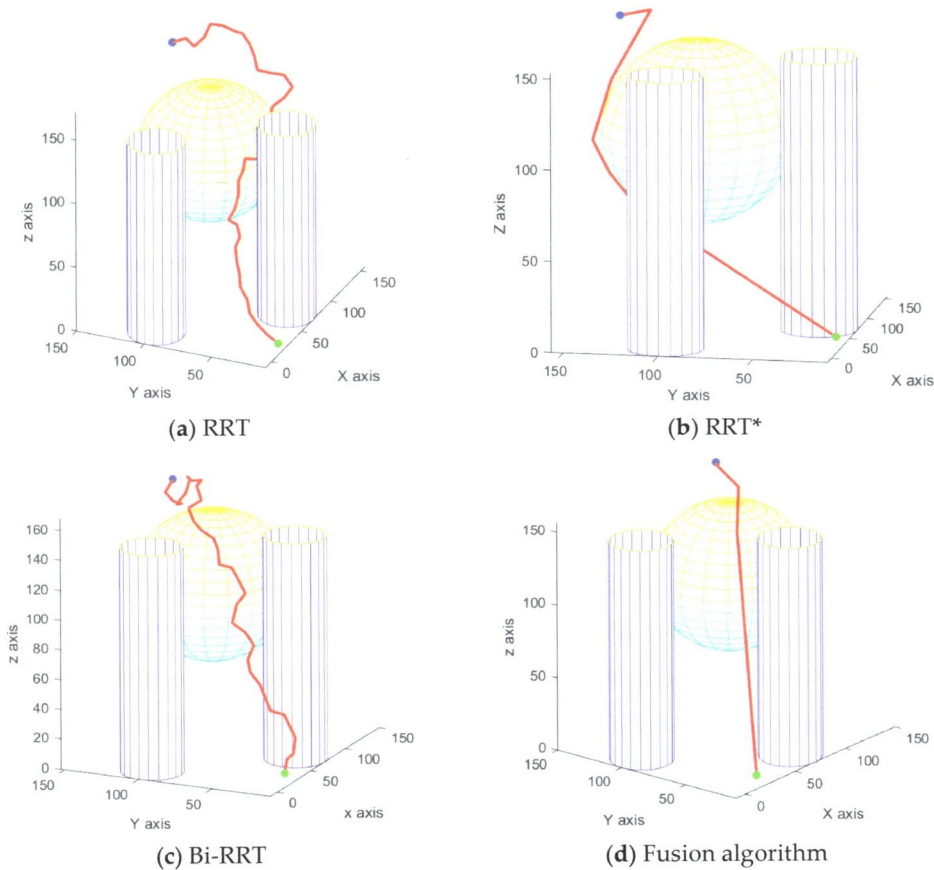

Figure 17. Different algorithm planning effect.

6.2. Simulation Validation

The experiments were performed using a virtual machine with the Ubuntu 18.04 operating system, with a melodic version of the ROS platform installed. For hardware, a PC (CPU i5-12600KF @3.70 GHz, RAM 32G, GPU RTX 4060TI 16G) was used to support the experiments.

The drilling and anchoring robot system model was established through Solidworks, which mainly includes the robotic arm with the material, a material magazine, a drilling machine, and a roof plate. It was imported into the ROS platform to build the working environment of the coal mine roadway excavation support. Simulate the trajectory planning algorithm in the ROS platform, write the corresponding code of the algorithm, and build the corresponding robot arm trajectory planning library. Keeping the rest of the parameters the same, simulate the support process of a drill hole and compare and analyze the planning results of different algorithms.

The following is a simplified step-by-step procedure for the drilling and anchoring robotic system to carry out the support work of coal mine tunnel excavation:

(1) Robotic arm for loading and unloading drill rods for drilling rig;
(2) Robotic arm facing the drilling hole and loading medical roll;
(3) Robotic arm facing the drilling rig for loading anchor rods.

The figure above shows the simulation results of robotic arm motion planning, in which the yellow surface represents the motion trajectory of the end of the robotic arm from the starting position to the final position. In the figure, the path planning carried out by the robotic arm in the state of clamping the drill pipe, the medicine roll, and the anchor, respectively, still maintains the vertical state of clamping objects, which is in line with the attitude requirements of the drilling and anchoring robotic system in carrying out the work of tunnel excavation support in coal mines. According to the trajectory diagrams from Figures 18–21, it can be found that the fusion algorithm plans the trajectory without any collision during the movement process; the overall process is smooth, and the trajectory is in the feasible domain of the robotic arm.

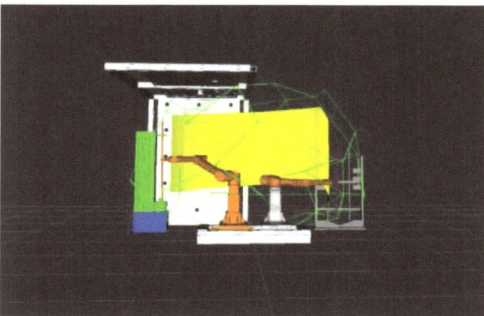

Figure 18. Grabbing the drilling rods and loading them into the drilling rig.

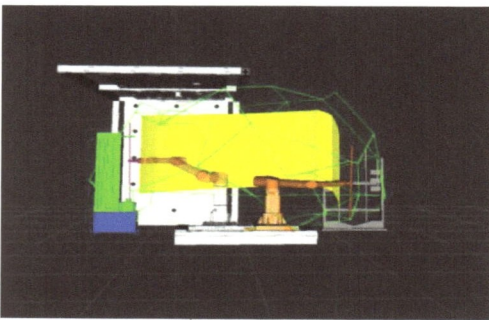

Figure 19. Remove the drill rod from the drilling rig and bring it back to the material depot.

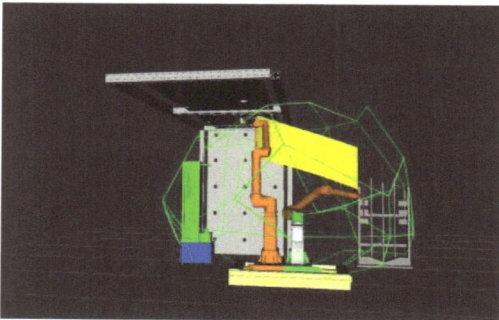

Figure 20. Clamping and loading of medical rolls into top plate drill holes.

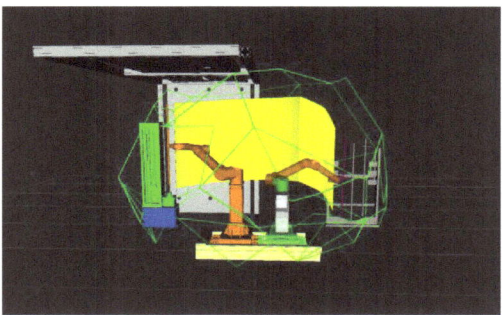

Figure 21. Grabbing anchor rods for loading into the drilling rig.

Table 3 shows the simulation data of different algorithms. Comparing with Table 3, it can be found that the longest planning time of the RRT algorithm is not ideal. Although Bi-RRT has a shorter time, the trajectory is longer, and there are more inflection points, and the RRT* algorithm has a short trajectory length due to its asymptotically optimal characteristics. However, the planning time of each time is up to the maximum, and this paper's algorithm is optimal in terms of both the speed and quality of the planning. The planning time is up to 69.8% compared to the RRT and up to 45.3% compared to Bi-RRT. The trajectory length is 29.8% compared to Bi-RRT and 13.4% compared to RRT*. The fusion algorithm has the highest success rate in the planning process, and the number of iterations is relatively shortest.

Table 3. Simulation data of different algorithms.

	Task 1					Task 2			
Algorithm	Planning Time/s	Trajectory Length/mm	Success Rate	Iteration Frequency	Algorithm	Planning Time/s	Trajectory Length/mm	Success Rate	Iteration Frequency
RRT	0.57	8770	56%	181	RRT	0.33	3580	62%	146
Bi-RRT	0.23	5352	78%	156	Bi-RRT	0.15	2782	86%	132
RRT*	5	5236	50%	124	RRT*	5	2460	66%	135
Ours	0.11	4050	100%	76	Ours	0.07	2320	100%	62
	Task 3					Total			
Algorithm	Planning Time/s	Trajectory Length/mm	Success Rate	Iteration Frequency	Algorithm	Planning Time/s	Trajectory Length/mm	Success Rate	Iteration Frequency
RRT	0.46	6850	64%	166	RRT	1.36	19,130	60.7%	164
Bi-RRT	0.37	6436	88%	168	Bi-RRT	0.75	14,570	84%	152
RRT*	5	4112	58%	148	RRT*	15	11,808	58%	135
Ours	0.23	3850	100%	102	Ours	0.41	10,220	100%	80

7. Conclusions

In the coal mine tunnel drilling and anchoring support process, when the robotic arm cooperates with the drilling machine to load and unload the drilling and anchoring materials, it is difficult to plan the trajectory in the dynamic environment. An obstacle avoidance trajectory planning algorithm based on the fusion of the improved artificial potential field and Bi-RRT is proposed. The global trajectory planning in a static environment is carried out by improving the Bi-RRT algorithm, and then the guiding potential field is established based on the optimized trajectory and real-time obstacle information; the final trajectory is derived using the improved artificial potential field method. Through Matlab simulation analysis, it is verified that the fusion algorithm has a better path-planning effect than the traditional algorithm and avoids the local minimum solution. The feasibility of the algorithm in the drilling and anchoring robot system is verified through the ROS platform Rviz simulation.

In order to further study the adaptability of the relevant algorithm in the complex environment of a coal mine underground, the drilling and anchoring robot system platform will be built in future work to simulate the real permanent support scene and improve the robustness of the algorithm.

Author Contributions: Conceptualization, S.S. and S.M.; methodology, S.S., S.M. and X.X.; software, C.W., X.X. and S.M.; validation, S.S., S.M. and H.Y.; formal analysis, S.S., C.W. and X.X.; investigation, H.M., Y.G. and H.Y.; resources, S.S., Y.G. and H.S.; data curation, H.S. and H.Y.; writing—original draft preparation, S.S., S.M. and H.Y.; writing—review and editing, S.S., S.M. and H.Y.; visualization, Y.G. and H.S.; supervision, X.X. and C.W.; project administration, X.X.; funding acquisition, X.X., C.W. and H.M. All authors have read and agreed to the published version of the manuscript.

Funding: This research was funded by the Key Technologies Research and Development Program of China under grant 2023YFC2907600, the National Natural Science Foundation of China project under grant 52374161, the Shaanxi Provincial Department of Education to Serve Local Special under grant 22JC051, and the Key Research and Development Projects of Shaanxi Province under grant 2023-LL-QY-03.

Institutional Review Board Statement: Not applicable.

Informed Consent Statement: Not applicable.

Data Availability Statement: The data presented in this study are available on request from the author.

Conflicts of Interest: The authors declare no conflicts of interest.

References

1. Li, X.; Cao, Z.; Xu, Y. Characteristics and trends of coal mine safety development. *Energy Sources* **2021**, 1–19. [CrossRef]
2. Wang, G.; Ren, H.; Zhao, G. Research and practice of intelligent coal mine technology systems in China. *Int. J. Coal Sci. Technol.* **2022**, *9*, 24. [CrossRef]
3. Hao, S.; He, T.; Ma, X.; Zhang, X.; Wu, Y.; Wang, H. KDBiDet: A Bi-Branch Collaborative Training Algorithm Based on Knowledge Distillation for Photovoltaic Hot-Spot Detection Systems. *IEEE Trans. Instrum. Meas.* **2024**, *73*, 3335509. [CrossRef]
4. Ren, B.; Li, L.; Yang, Y. Research on an Intelligent Mining Complete System of a Fully Mechanized Mining Face in Thin Coal Seam. *Sensors* **2023**, *23*, 9034. [CrossRef]
5. Kang, H.; Jiang, P.; Song, D.; Lei, Y. Intelligent rapid drilling equipment with integrated drilling and anchoring. *Intell. Min.* **2023**, *4*, 9–14.
6. Liu, K.; Cheng, J.; Sun, X.; Li, Z. Automated Stratum Interface Detection Using the Optimized Drilling Specific Energy through Self-Adaptive Logistic Function. *Sensors* **2023**, *23*, 8594. [CrossRef]
7. Hao, S.; An, B.; Ma, X.; Sun, X.; He, T.; Sun, S. PKAMNet: A Transmission Line Insulator Parallel-Gap Fault Detection Network Based on Prior Knowledge Transfer and Attention Mechanism. *IEEE Trans. Power Deliv.* **2023**, *38*, 3387–3397. [CrossRef]
8. Li, K.; Yao, Y. Research on key technology of automatic loading and unloading of dual pipe directional drilling. *Vibroeng. Procedia* **2021**, *38*, 161–165.
9. Zhang, J.; Wang, Y.; Che, L.; Wang, N. Workspace analysis and motion control strategy of robotic mine anchor drilling truck manipulator based on the WOA-FOPID algorithm. *Front. Earth Sci.* **2022**, *10*, 954547.
10. Li, D.; Wang, T.; Du, H. Research on trajectory tracking of automatic loading and unloading drill pipe robotic arm based on fuzzy sliding mode backstepping control. *J. Nanjing Univ. Sci. Technol.* **2023**, *47*, 619–628.
11. Ma, H.; Sun, S.; Wang, C. Key technology of coal mine roadway drilling and anchoring robot with multi-mechanical arm and multi-drilling machine collaboration. *J. China Coal Soc.* **2023**, *48*, 497–509.
12. Chen, D.; Tan, Q.; Xu, Z. Robotic arm path planning based on sampling point optimization RRT algorithm. *Control. Decis. Mak.* **2024**, *39*, 2597–2604.
13. Xiao, Q.; Xiang, G.; Chen, Y.; Zhu, Y.; Dian, S. Time-Optimal Trajectory Planning of Flexible Manipulator Moving along Multi-Constraint Continuous Path and Avoiding Obstacles. *Processes* **2023**, *11*, 254. [CrossRef]
14. Zhang, N.; Cui, C.; Wu, G. Path planning of a 5-dof robotic arm based on BiRRT-APF algorithm considering obstacle avoidance. *Proc. Inst. Mech. Eng. Part C J. Mech. Eng. Sci.* **2022**, *236*, 9282–9292. [CrossRef]
15. Wang, G.; Ku, X.; Wu, H. Obstacle avoidance path planning for 6R robotic arm based on improved APF-RRT. *Mach. Tools Hydraul.* **2024**, *52*, 27–32.
16. Chen, G.; Liu, J. Mobile Robot Path Planning Using Ant Colony Algorithm and Improved Potential Field Method. *Comput. Intell. Neurosci.* **2019**, *2019*, 1932812. [CrossRef]

17. Chen, C.; Huang, Z.; Tseng, D. Biased-RRT correction algorithm for six-degree-of-freedom robotic arm path planning in complex environments. *J. Fuzhou Univ. (Nat. Sci. Ed.)* **2022**, *50*, 658–666.
18. Zhang, Y.; Shi, G.; Xu, J. Surface unmanned craft path planning algorithm based on artificial potential field method guided Bi-RRT. *J. Shanghai Marit. Univ.* **2022**, *43*, 16–22.
19. Friston, S.; Steed, A. Real-time collision detection for deformable characters with radial fields. *IEEE Trans. Vis. Comput. Graph.* **2019**, *25*, 2611–2622. [CrossRef]
20. Zhao, Z.; Liu, Z. Research on interference judgment of industrial robots based on spatial position. *Microprocessor* **2018**, *39*, 53–60.
21. Ma, Y.; Deng, Y.; Lai, M. Singular avoidance trajectory planning method for industrial robots. *Tool Technol.* **2024**, *58*, 144–150.
22. Feng, Y.; Zhao, Z.; Yan, J. Improved RRT algorithm for quadcopter UAV path planning. *J. Shenyang Univ. Technol.* **2024**, *43*, 9–15.
23. Wang, S. Obstacle avoidance control method of unloading arm moving process based on artificial potential field method. *Chem. Manag.* **2023**, *34*, 121–124.
24. Song, B.; Miao, H.; Xu, L. Path planning for coal mine robot via improved ant colony optimization algorithm. *Syst. Sci. Control Eng.* **2021**, *9*, 283–289. [CrossRef]
25. Xue, G.; Wang, Z.; Wang, Y. Path planning for underground coal mine robots based on improved artificial potential field algorithm. *Ind. Min. Autom.* **2024**, *50*, 6–13.

Disclaimer/Publisher's Note: The statements, opinions and data contained in all publications are solely those of the individual author(s) and contributor(s) and not of MDPI and/or the editor(s). MDPI and/or the editor(s) disclaim responsibility for any injury to people or property resulting from any ideas, methods, instructions or products referred to in the content.

Article

RSDNet: A New Multiscale Rail Surface Defect Detection Model

Jingyi Du, Ruibo Zhang *, Rui Gao, Lei Nan and Yifan Bao

College of Electrical and Control Engineering, Xi'an University of Science and Technology, Xi'an 710054, China; 000248@xust.edu.cn (J.D.); gaorui@xust.edu.cn (R.G.); 22206223076@stu.xust.edu.cn (L.N.); 23206223101@stu.xust.edu.cn (Y.B.)
* Correspondence: 22206223068@stu.xust.edu.cn

Abstract: The rapid and accurate identification of rail surface defects is critical to the maintenance and operational safety of the rail. For the problems of large-scale differences in rail surface defects and many small-scale defects, this paper proposes a rail surface defect detection algorithm, RSDNet (Rail Surface Defect Detection Net), with YOLOv8n as the baseline model. Firstly, the CDConv (Cascade Dilated Convolution) module is designed to realize multi-scale convolution by cascading the cavity convolution with different cavity rates. The CDConv is embedded into the backbone network to gather earlier defect local characteristics and contextual data. Secondly, the feature fusion method of Head is optimized based on BiFPN (Bi-directional Feature Pyramids Network) to fuse more layers of feature information and improve the utilization of original information. Finally, the EMA (Efficient Multi-Scale Attention) attention module is introduced to enhance the network's attention to defect information. The experiments are conducted on the RSDDs dataset, and the experimental results show that the RSDNet algorithm achieves a mAP of 95.4% for rail surface defect detection, which is 4.6% higher than the original YOLOv8n. This study provides an effective technical means for rail surface defect detection that has certain engineering applications.

Keywords: rail surface defect detection; YOLOv8; CDConv; BiFPN; EMA

Citation: Du, J.; Zhang, R.; Gao, R.; Nan, L.; Bao, Y. RSDNet: A New Multiscale Rail Surface Defect Detection Model. *Sensors* **2024**, *24*, 3579. https://doi.org/10.3390/s24113579

Academic Editor: Mario Luca Fravolini

Received: 8 May 2024
Revised: 27 May 2024
Accepted: 29 May 2024
Published: 1 June 2024

Copyright: © 2024 by the authors. Licensee MDPI, Basel, Switzerland. This article is an open access article distributed under the terms and conditions of the Creative Commons Attribution (CC BY) license (https://creativecommons.org/licenses/by/4.0/).

1. Introduction

During the train operation, frequent collisions between wheels and rails and the factor of outdoor environmental erosion can easily lead to defects on the rail surface, which may cause serious accidents if not handled in time [1]. The initial stage of track defect detection mainly adopts the inspection method; however, this method is inefficient and easily affected by subjective factors [2]. With the development of sensor and communication technologies, fault detection based on the dynamic response of wheels and rails has been widely used. Rail inspection has shifted to the use of sensors and automated equipment such as ultrasonic detection and eddy current detection [3–7]. This method uses sensors to capture vibration and acceleration signals during operation and analyzes these signals to identify abnormalities in the wheel-rail system and determine faults [8]. Fu et al. [9] simulate flatness anomalies using the multi-body dynamics software SIMPACK and generate spectral images for anomaly detection by analyzing acceleration signals. Xie et al. [10] developed a vehicle-track coupled dynamics model to simulate the dynamic response of the axle box under different speeds and track wear excitations. The sensor-based method has low environmental requirements and a small cost [11–13]. However, it may not be sensitive enough to detect subtle faults in some cases (e.g., surface cracks or minor wear), and it also affects the accuracy of detection in complex environments, such as when there is a large amount of noise interference. At the same time, the sensors mainly detect vibration and do not provide a visual image of the fault, which is not intuitive enough in some cases where quick diagnosis and repair are required.

In recent years, machine vision detection has been widely used in rail surface defect detection due to its accurate, rapid, and non-contact characteristics [14]. Machine vision detection is categorized into traditional image processing methods and deep learning detection methods as per the development time [15]. Traditional image processing methods require manually designed features or predefined defect features. The defects in the image are identified and localized by classifier settings, morphological operations, etc. [16]. Gan et al. [17] localized the railroad surface image defects through a two-stage algorithm, where the rough extractor initially locates the defects in the railroad surface image and the detail extractor further determines whether the anomalies are real defects or not. Zhang et al. [18] used a curvature filter to extract the structural information of the rail surface and used an improved Gaussian mixture model to identify the defects. Nieniewski [19] employed morphological operations such as corrosion and expansion to highlight defective features on the rail surface and identified defective areas in the image through setting thresholds and conditions. However, traditional machine learning detection methods are sensitive to noise in images and have poor generalization ability, which limits their application in real-rail defect detection.

With the rapid development of deep learning, convolutional neural networks have become an obvious choice for defect detection on rail surfaces due to their unique feature representation advantages and modeling capabilities. Based on the utilization of region proposal networks, deep learning-based object detection methods are classified into two-stage or one-stage networks [20]. Typical algorithms for two-stage target measurement include R-CNN [21], Faster-RCNN [22], and Mask-RCNN [23], etc. These algorithms utilize RPN to quickly generate and screen out candidate regions containing rail surface defects at the initial stage, providing a basis for subsequent defect classification and localization. Yu et al. [24] proposed the method of migration learning to train the network to realize rail surface defect detection. Bai et al. [25] used faster R-CNN to classify and detect the labeled image dataset and enhanced the detection rate and accuracy by optimizing the anchor box function. Wang et al. [26] designed a multi-scale feature pyramid based on a two-stage network to adapt to the detection of track defects of different sizes, and the CIOU evaluation metrics were also introduced to optimize the performance of the RPN to achieve the precise location of the defects. Although the two-stage inspection method has advantages in accuracy, it still faces challenges such as inaccurate candidate region generation, slow detection speed, and difficulty in recognizing small-scale defects in practical applications.

One-stage target detection methods use an end-to-end approach to accomplish the target detection task directly without generating candidate boxes, and the typical algorithms for single-stage target measurement are the YOLO (You Only Look Once) series [27–32], SSD (Single Shot MultiBox Detector) [33], etc., which have the advantages of simplicity and high efficiency, etc. [34]. The YOLO series offers significant advantages in terms of fast response, efficient deployment, and adaptability, and more and more researchers are using YOLO algorithms in rail surface defect detection. Wang et al. [35] designed spatial attention sharpening filters based on YOLOv5s to enhance attention to the defects at the edge location of the rail defects and constructed M-ASFF to enhance the details of the underlying features of tiny defects. Zhang et al. [36] used BiFPN for feature fusion at the neck of YOLOX and also fused the NAM attention mechanism to improve the image feature expression ability, and the experimental results showed that the defect recognition rate was improved by 2.42% compared to YOLOX. Wang et al. [37] addressed the problem of detecting small targets and dense occlusion on the surface of rails by introducing the SPD-Conv building block in YOLOv8 to improve detection attention to small and medium-sized targets, and the Focal-SIoU loss function was used to adjust the sample weights to improve the model's ability to recognize complex samples. The YOLOv8 network is one of the newest open-source neural networks in the YOLO family, offering high performance in terms of detection speed and accuracy [38].

In conclusion, to address the issue of rail surface defects with different scales and many small-scale defects, this study proposes a track surface defect detection algorithm, RSDNet, based on the improved YOLOv8 algorithm. The primary contributions of this study are summarized as follows:

(1) Proposed CDConv (Cascade Dilated Convolution), a module based on feature reuse. It was introduced into Backbone to realize multi-scale feature extraction without increasing the number of too many parameters.
(2) Based on the idea of BiFPN (Bi-directional Feature Pyramids Network), change the feature fusion method of Head, add jump connections, and utilize more original feature information for feature fusion to improve the network's ability to recognize defective edges.
(3) Incorporate the EMA (Efficient Multi-Scale Attention) module into Head to enhance the feature extraction network's attention to defect detail information, thus improving the detection accuracy of rail surface defects.

2. YOLOv8

YOLOv8 is a one-stage target detection algorithm proposed by Ultralytics in 2023. Its performance is so superior that it outperforms most of the target detection algorithms. Therefore, YOLOv8 is the baseline model chosen in this study. YOLOv8 is mainly composed of the Backbone, Head, and Detector, and its structure is shown in Figure 1.

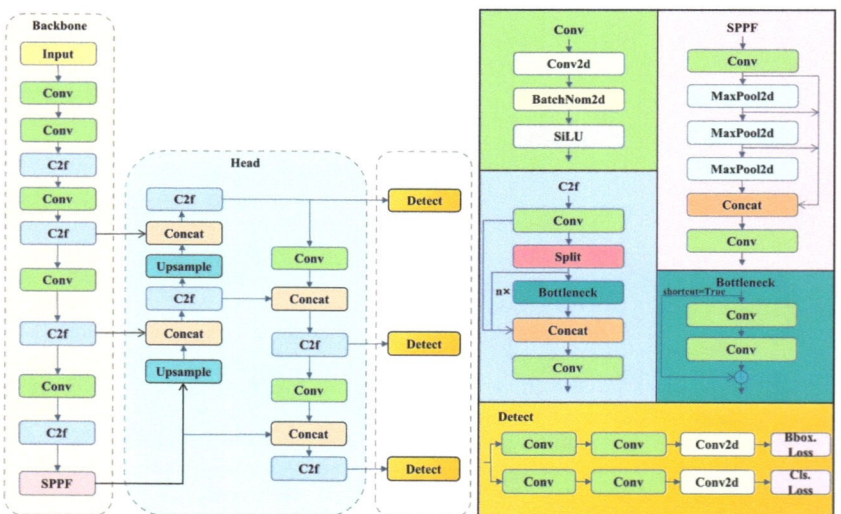

Figure 1. The structure of YOLOv8.

YOLOv8 adjusts the input image to 640 × 640 resolution. The Backbone consists of CBS, C2f, and SPPF modules. The CBS module includes Conv, BatchNormal, and SiLU, which realize the transformation and extraction of the input features; the C2f module captures the gradient flow information by using Bottleneck units; and the SPPF module reduces the computation and improves the feature extraction efficiency through the serial stacked pooling layer to reduce the computation and enhance the feature extraction efficiency. Head adopts the PANet structure to realize feature fusion and information transfer. Detect uses a decoupled Head to separate the regression and prediction branches. Through the DFL (Distribution Fusion Loss) strategy, the regression coordinates are regarded as distributions rather than single values, which helps the model deal with the defects of small scales or irregular shapes in a way that provides more accurate localization information.

YOLOv8n, as the smallest model in the YOLOv8 series, has the advantages of fast detection speed and low resource consumption. However, if YOLOv8n is directly applied to the task of detecting defects on the rail surface, the model will face challenges such as diverse defect scales and more small-scale defects. In order to solve these problems, targeted adjustments to the model are needed to enhance the model's ability to recognize targets at different scales, thereby improving the overall performance of track surface defect detection.

3. RSDNet: YOLOv8n-CDConv-BiFPN-EMA

RSDNet is based on the YOLOv8n model, and the designed CDConv module is introduced in Backbone to realize multi-scale feature extraction. Drawing on the BiFPN idea in Head, a new fusion method is designed to enhance the network's utilization of raw information. Meanwhile, EMA is fused to enhance attention to the defect information. Figure 2 illustrates the architectural design of the RSDNet model. Where ① denotes the designed CDConv module, ② denotes the designed feature fusion method, and ③ denotes the location where EMA is added.

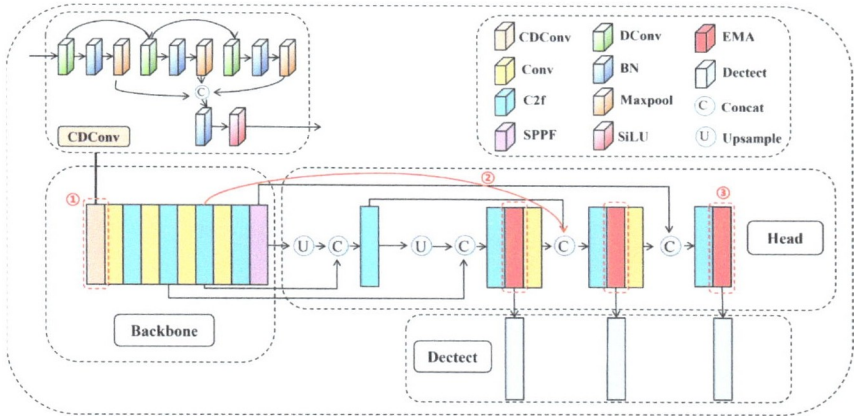

Figure 2. The structure of the proposed method, RSDNet. (YOLOv8n-CDConv-BiFPN-EMA).

Compared with YOLOv8, the improved algorithm is able to capture long-distance dependencies in the image earlier, fuse more low-level semantic information such as defect edges, and dynamically emphasize and guide key defect detail information to achieve more accurate rail surface defect detection.

3.1. CDConv Module Proposed in This Study

Dilated convolution is a convolution technique proposed by Google in 2015 for enlarging the receptive field, which was first applied to the DeepLab model [39]. Dilated convolution is able to increase the receptive field without increasing the number of parameters as shown in Figure 3, the same 3×3 convolution kernel can have the effect of 5×5 and 7×7 convolution.

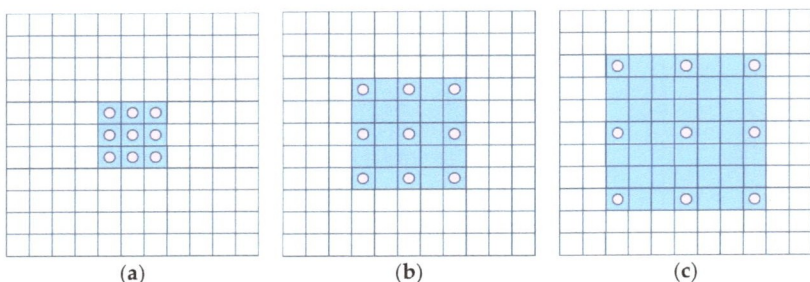

Figure 3. Comparison between Regular Convolution and Dilated Convolution. (**a**) is a regular convolution process (dilation rate = 1), and the receptive field is 3; (**b**) is the dilated convolution with dilation rate = 2, and the receptive field is 5; (**c**) is the dilated convolution with dilation rate = 3, and the receptive field is 7.

When detecting defects on the rail surface, the direct use of multiple cavity convolutions tends to increase too many parameters, although it can increase the receptive field. The CDConv module designed in this paper adopts a feature reuse strategy to realize parameter sharing by connecting multiple cavity convolution layers in series, which reduces the consumption of computational resources. In addition, the CDConv module realizes multi-scale feature extraction of the input defective image by setting multiple parallel branches with different receptive fields. The CDConv module is shown in Figure 4.

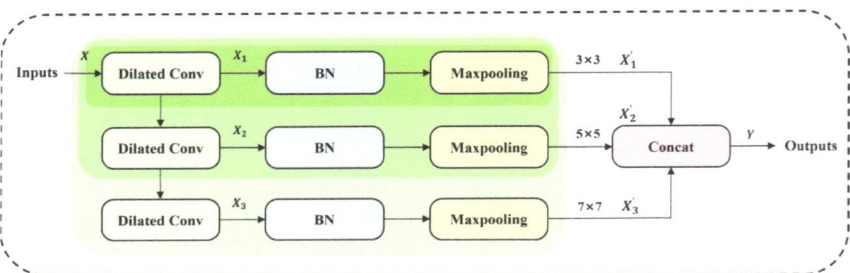

Figure 4. The structure of the Cascaded Dilated Convolution (CDConv).

Assume the input feature mapping is X. With three cascading null convolutions, it can be obtained as:

$$X_1 = D_1(X), \tag{1}$$

$$X_2 = D_2(X_1), \tag{2}$$

$$X_3 = D_3(X_2), \tag{3}$$

where $D_i(\cdot)$ is a dilated convolution operation with a specific dilated rate.

Next, the output of each dilated convolution is subjected to BatchNormal (BN) and Maximum Pooling ($Maxpooling$) layers to further optimize the feature representation and speed up the computational process. These operations can be represented as:

$$X'_1 = Max(BN(X_1)), \tag{4}$$

$$X'_2 = Max(BN(X_2)), \tag{5}$$

$$X'_3 = Max(BN(X_3)), \quad (6)$$

where $BN(\cdot)$ and $Max(\cdot)$ denote the operations of BN and $Maxpooling$, respectively, and X'_i are the outputs after BatchNormal and Maximum Pooling.

By connecting these three output feature maps, a feature representation containing multi-scale information can be obtained.

$$Y = [X'_1, X'_2, X'_3], \quad (7)$$

Aiming at the problem that small target defects occupy few pixels in the image, this study applies CDConv to the first layer after the input of the YOLOv8n model so that the model can maintain a higher resolution from the beginning, and improve the model's ability to recognize and localize the small target defects in the initial stage. Meanwhile, the flexible feature extraction capability of CDConv can better adapt to various shape defects that may appear in complex scenes, making the model more robust.

3.2. Feature Extraction Method in This Study

BiFPN is a weighted bidirectional feature pyramid network proposed by Mingxing Tan et al. in EffiicientDet [40]. BiFPN facilitates the flexibility and effectiveness of the information flow of feature maps between different layers by introducing weighted bidirectional connections. Compared to the fusion approach originally adopted by YOLOv8n, BiFPN realizes cross-scale connectivity with the design changes shown in Figure 5.

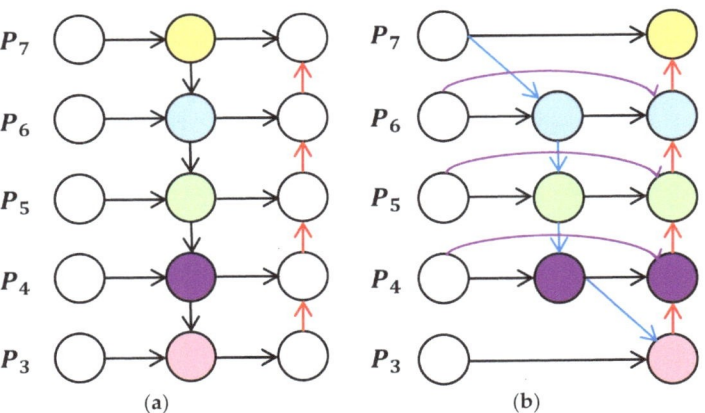

Figure 5. Feature network design. (**a**) PANet; (**b**) BiFPN.

In the design of the YOLOv8n network, the feature fusion strategy employs an optimized PANet structure to enhance feature integration efficiency. However, this fusion method fails to fully exploit the potential of the original feature information. To further improve the performance of the model, this study draws on the idea of BiFPN to improve the feature fusion mechanism.

The improved feature fusion is shown in Figure 6. After the third C2f layer of the backbone, a new path is introduced to fuse the features extracted from this layer with the features from the first C2f layer of the Head part and the first Conv layer of the neck part. This design makes the raw, detail-rich features that have not been multiprocessed participate more in the feature fusion process, reduces the loss of information in the transfer process, and makes the model capture the defect detail information more acutely.

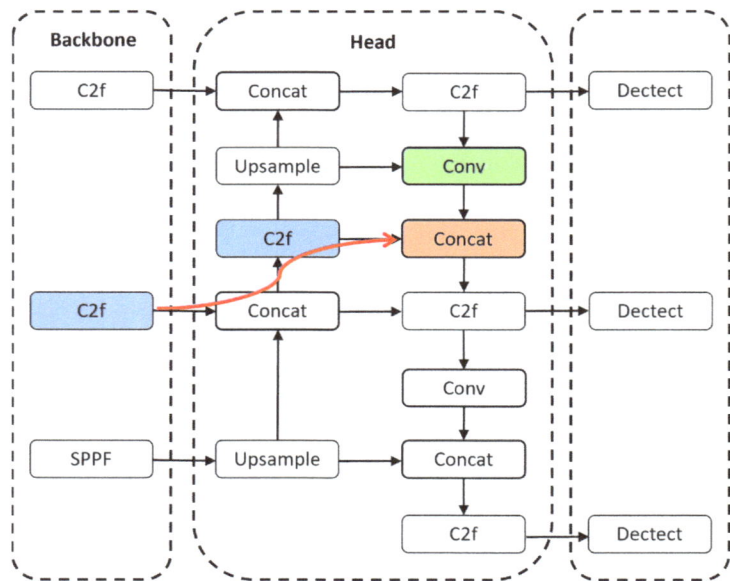

Figure 6. The structure of feature fusion.

Since different input features have different resolutions, they usually contribute unequally to the output features. To solve this problem, BiFPN adds an extra weight to each input and allows the network to learn the importance of each input feature. Normalized fusion is shown in Equation (8), which is less computationally intensive and has similar accuracy compared to Softmax function-based fusion methods.

$$OUT = \sum_i \frac{w_i}{\varepsilon + \sum_j w_j} \cdot IN_i, \qquad (8)$$

where w_i is a learnable weight that can be a scalar (per feature), a vector (per channel), or a multidimensional tensor (per pixel).

By fusing more feature information and improving the utilization of the original feature information, the generalization ability of the model and the recognition ability of the defect edge information are improved. The weighting mechanism also ensures that the importance of different levels of features is reasonably balanced, which helps to improve the model's ability to detect defects in small targets.

3.3. Head Network with EMA

Due to the complex and variable background of the track surface and the difficulty in detecting subtle defects, the EMA attention mechanism is introduced in order to further enhance the screening and filtering abilities of the network on key information and improve the performance of the algorithm. EMA is a cross-space learning approach proposed by Daliang Ouyang et al., Efficient Multi-Scale Attention, that can interact with information without channel dimensionality reduction and reduce computational overhead [41]. Its structure is shown in Figure 7. In this image + indicates an addition operation and ∗ indicates a multiplication operation.

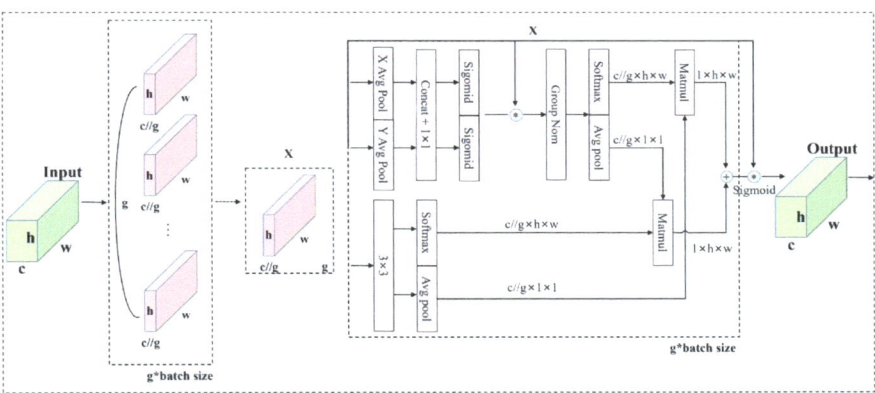

Figure 7. EMA mechanism structure diagram.

The EMA module divides the input feature map X along the channel dimension into G groups, each of which can be represented as X_i. Each sub-feature group is learned to obtain the corresponding weights, allowing the network to focus on different regions and features in the track surface image.

$$X_i = \{X_{i,j}\}_{j=1}^{C/G}, \tag{9}$$

where i is the index of the group, C is the total number of input channels, and G is the number of subgroups.

For each group X_i, EMA employs two parallel branches to capture cross-dimensional interactions capturing pixel-level relationships, and improved feature representation, the outputs of which can be denoted as $F_{1\times1}$ and $F_{3\times3}$. For the outputs of the 1×1 branch and the 3×3 branch, the channel weights are adjusted using 2D global average pooling coding, respectively.

$$P_{1\times1} = \frac{1}{H \times W} \sum_{i=1}^{H} \sum_{j=1}^{W} F_{1\times1}(i,j), \tag{10}$$

$$P_{3\times3} = \frac{1}{H \times W} \sum_{i=1}^{H} \sum_{j=1}^{W} F_{3\times3}(i,j), \tag{11}$$

Finally, the information from these two branches is fused by a matrix dot product operation to get the final output feature map X_{EMA}.

$$X_{EMA} = \sigma(P_{1\times1} \cdot P_{3\times3} + X), \tag{12}$$

where σ is the Sigmoid activation function,· denotes the matrix dot product operation, and X is the original input feature map. The matrix operation captures pixel-level relationships while avoiding the lack of channel information richness due to dimensionality reduction.

As shown in Figure 8, the EMA is combined with the three C2f connecting Detect in the Head. The EMA combines the C2f to ensure the full utilization of the features at each scale. The cross-space learning mechanism of the EMA is able to aggregate defect feature information from different branches. At the same time, EMA dynamically adjusts the weights in such a way that it can strengthen the key defect regions in the track surface feature map and ignore irrelevant background information, so as to improve the accuracy of track surface defect detection.

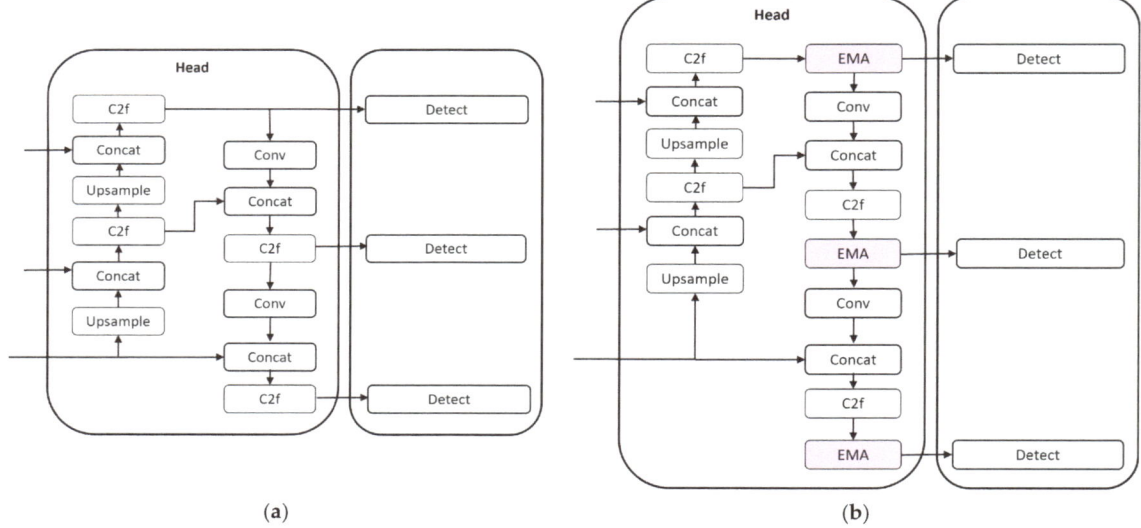

Figure 8. Schematic of the location where the EMA module is added. (**a**) The structure of the Head; (**b**) The structure of added EMA Head.

4. Experiments

4.1. Experimental Details

4.1.1. Dataset

In this study, the RSDDs dataset collected and made publically available by Beijing Jiaotong University was used, as shown in Figure 9. The dataset contains 195 difficult rail surface defect photos, with 67 being 160 pixels × 1000 pixels and 128 being 55 pixels × 1250 pixels. After segmentation and image adjustment screening, 240 track surface defect images were obtained, and an initial track surface defect image sample dataset was established. Then, based on the initially established dataset, the dataset is expanded using the flip transform, random cropping, and brightness transform. Finally, 1080 rail surface defect images are generated, in which the ratio of the training set, validation set, and test set is 7:2:1.

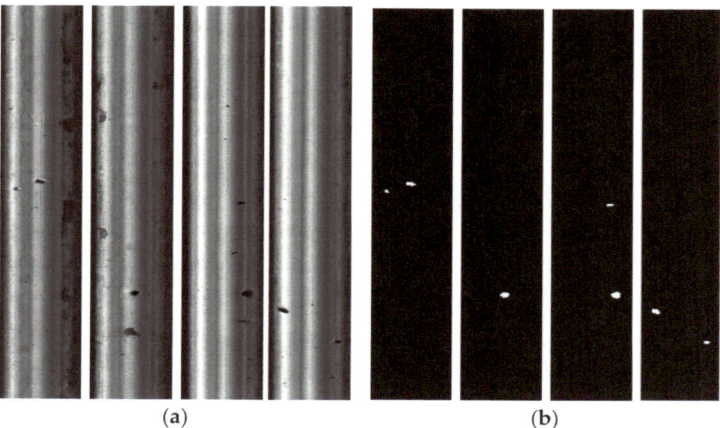

Figure 9. Examples of RSDD datasets. (**a**) Rail surface images; (**b**) GroundTruth.

4.1.2. Experimental Environment

Experimental and Parameter Configuration The experimental setup is shown in Table 1. For model training, this study used an Intel(R) Xeon(R) CPU E5-2640 v4 (Intel Corporation, Santa Clara, CA, USA) and an NVIDIA Quadro P2200 GPU (NVIDIA Corporation, Santa Clara, CA, USA). The software environment consisted of CUDA version 10.2, Python 3.8, and Pytorch version 2.2.1.

Table 1. Experimental environment configuration.

Experimental Component	Version
CPU	Intel(R) Xeon(R) CPU E5-2640 v4
GPU	NVIDIA Quadro P2200
CUDA version	10.2
Python version	3.8
Pytorch version	2.2.1

4.1.3. Evaluation Index

In this study, precision (P), recall (R), and mean average precision (mAP) were used as metrics to assess the effectiveness of the algorithm for detection. Among them, mAP comprehensively evaluates the performance of the model in detecting all categories. The following are the specific calculations of the evaluation metrics:

$$P = \frac{TP}{TP + FP}, \tag{13}$$

$$R = \frac{TP}{TP + FN}, \tag{14}$$

where P denotes the prediction precision of the model and R denotes the recall of the model. TP represents the number of correctly classified positive samples, FP represents the number of misclassified positive samples and FN represents the number of misclassified negative samples.

$$AP = \int_0^1 P(R)dR, \tag{15}$$

$$mAP = \frac{1}{n}\sum_{i=1}^{n} AP_i, \tag{16}$$

where AP represents the area under the precision-recall curve for a particular category at different confidence thresholds. mAP stands for mean accuracy and is the average of the APs for each category.

4.2. Results and Discussion

4.2.1. Ablation Experiments

In this paper, ablation experiments of CDConv, BiFPN, and EMA attention mechanisms were performed sequentially. The results of the experiments are shown in Table 2.

Table 2. Comparison of results of ablation experiments.

Model	Parameter (M)	P (%)	R (%)	mAP@0.5 (%)
YOLOv8n	**30.06**	92.3	86.7	90.8
YOLOv8n+CDConv	30.11	96.0	88.6	94.1
YOLOv8n+BiFPN	30.22	95.6	87.3	92.3
YOLOv8n+EMA	30.17	95.8	87.9	93.7
YOLOv8n+CDConv+BiFPN	30.28	96.2	88.5	94.8
YOLOv8n+BiFPN+EMA	30.36	95.9	88.3	94.4
YOLOv8n+CDConv+BiFPN+EMA	30.41	**96.4**	**90.6**	**95.4**

The first row of data in Table 2 shows the performance of the original YOLOv8n model without any improvements, where the average accuracy on the rail surface defect detection task is 90.8%. To further improve the performance of the model, the convolution operation for the Backbone and the fusion method for the Head have been improved. In the second and third rows of the table, the model performance after introducing the CDConv module and adopting the BiFPN fusion method is shown, respectively. By adding the CDConv module, the model's mAP on rail face defect detection is improved to 94.1%, and the multi-scale feature extraction capability of CDConvd enhances the model's detection capability. The mAP also reaches 92.3% with BiFPN fusion, which is due to BiFPN's ability to integrate global and contextual information more effectively. Finally, the EMA attention mechanism is introduced in the Head section, which significantly improves the accuracy of the model, even though this improvement leads to only a slight increase in the number of parameters. The EMA module further enhances the detection performance of the model by focusing on hinge defect features.

The mAP is the area enclosed by the mapping of precision and recall on the two axes. In the experiments in this paper, the mAP curves of the initial YOLOv8n model and the improved YOLOv8n model are shown in Figure 10. These curves depict the performance trend of the two models during the training process. As can be seen from the figure, the mAP curves of the improved YOLOv8n show a clear upward trend throughout the training process, which indicates that the performance of the model is steadily improving with the training. After sufficient training, the mAP curve finally stabilizes at a higher plateau, indicating that the model's performance has reached its peak. The mAP of the original YOLOv8 is 90.8%, while the mAP of this paper is 95.4%, which is 4.6% higher.

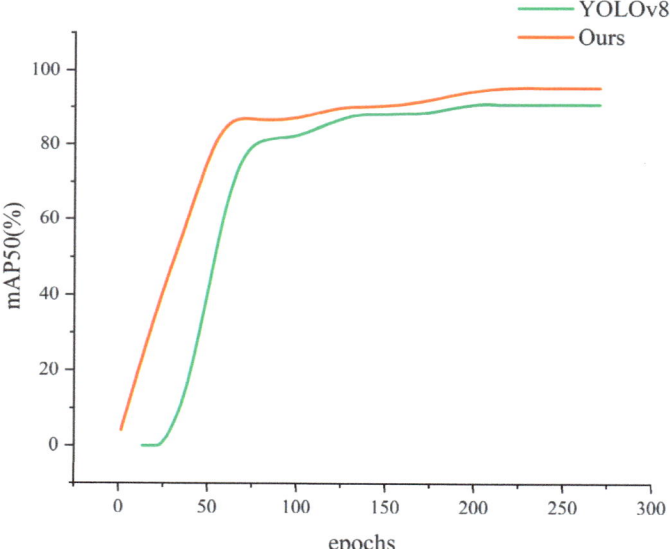

Figure 10. The mAP curves for the original YOLOv8 and the RSDNet.

4.2.2. Improved Model Comparison Experiments

This section experiments with CDConv, EMA position, and quantity in turn.

- Location and number of CDConv

Table 3 represents the effect of changing different numbers of Conv to CDConv at a variety of locations in the network model on the experimental results.

Table 3. Effect of the position and number of CDConv on the model.

Model	Position	Numbers	P (%)	R (%)	mAP@0.5 (%)
YOLOv8n+CDConv	Backbone	1	95.4	88.9	94.1
YOLOv8n+CDConv	Backbone	3	94.7	88.4	93.9
YOLOv8n+CDConv	Backbone	5	95.2	88.6	91.4
YOLOv8n+CDConv	Head	1	94.3	88.6	88.8
YOLOv8n+CDConv	Head	2	94.4	88.5	89.4

As can be seen from Table 3, adding CDConv modules to the backbone part of the network can effectively improve the detection performance of the model, especially when adding 1 module. With the increase in the number of CDConv modules, although the recall of the model has improved to some extent, accuracy and mAP@0.5 growth trends have leveled off or even decreased, which may be due to the overfitting caused by the increase in the model complexity. Adding CDConv modules to the Head part of the network also helps to improve the model performance, but the improvement is more limited compared to the backbone part. With this in mind, the first convolutional layer of Backbone is replaced with CDConv in this paper.

- Location and number of the EMA

Table 4 represents the effect of adding different numbers of EMAs at a variety of locations in the network model on the experimental results.

Table 4. Effect of the position and number of EMA on the model.

Model	Position	Numbers	P (%)	R (%)	mAP@0.5 (%)
YOLOv8n+EMA	Backbone	1	94.2	86.5	90.9
YOLOv8n+EMA	Backbone	3	95.4	87.1	91.9
YOLOv8n+EMA	Backbone	4	95.2	87.6	92.5
YOLOv8n+EMA	Head	3	95.8	87.9	93.7
YOLOv8n+EMA	Head	4	94.6	87.5	93.2

From Table 4, it can be seen that although EMA can effectively improve the accuracy of the backbone network part, the improvement is not as obvious as that of the Head. This may be related to the role of EMA in different network layers; Head, as the feature fusion layer in the model, serves to integrate the feature maps of different layers to provide rich contextual information for Dectect, while applying EMA in Head is more helpful for smoothing and stabilizing these features. Considering this, this paper adds EMA modules after the three C2f's connected to Dectect in the Head.

4.2.3. Performance Comparison Experiments

The performance of the improved algorithm is compared with several other typical target detection models in this work.

As shown in Table 5, the algorithm significantly outperforms the traditional Faster R-CNN and SSD algorithms for rail surface defect detection. Compared with the YOLOv5s, YOLOv7-Tiny, and YOLOv8n models in the YOLO family, the improved algorithm achieves 5.6%, 7.4%, and 4.6% on mAP@0.5. The inference time is 2.5 ms faster than YOLOv5s, but 0.4 ms and 1.3 ms slower than YOLOv7-Tiny and YOLOv8n, respectively. In addition, the algorithm also demonstrates high levels of accuracy and recall, two key performance metrics. These results fully validate the accuracy and stability of the RSDNet model in the task of rail surface defect recognition, which meets the practical needs of rail inspection.

Table 5. The results of comparison experiments.

Model	P (%)	R (%)	mAP@0.5 (%)	Times/ms
Faster R-CNN	71.5	72.9	72.3	156.5
SSD	78.3	67.0	75.1	46.2
YOLOv5s	86.7	80.8	89.8	22.3
YOLOv7-tiny	85.7	83.3	88.0	19.4
YOLOv8n	92.3	87.7	90.8	18.5
Ours	**96.4**	**90.6**	**95.4**	19.8

Figure 11 compares the performance of YOLOv5, YOLOv7-Tiny, YOLOv8n, Faster R-CNN, SSD, and optimized YOLOv8n for defect detection in this paper. As can be seen from the figure, Faster R-CNN and SDD have missed detection when the defective image is small, as shown in the first and third rows of the example in Figure 11, where the two algorithms fail to detect the tiny defects located in the lower-left corner of the image. Although YOLOv5s and YOLOv7-tiny could recognize most of the defects, they suffer from low detection accuracy, as shown in the second row of Figure 11, with a detection value of less than 0.75. Also miss-detection occurs when the defects are located in a complex background environment. This may be because these models are designed with limited detection capability for small targets or because they are not robust enough when dealing with complex backgrounds. In contrast, the optimized YOLOv8n algorithm proposed in this paper performs well in the defect detection task, not only capturing all defects in the image comprehensively but also significantly improving the accuracy of the bounding box prediction. This indicates that the optimized YOLOv8n improves the ability to identify and locate defects through algorithmic improvements while maintaining the fast processing of the YOLO series of algorithms.

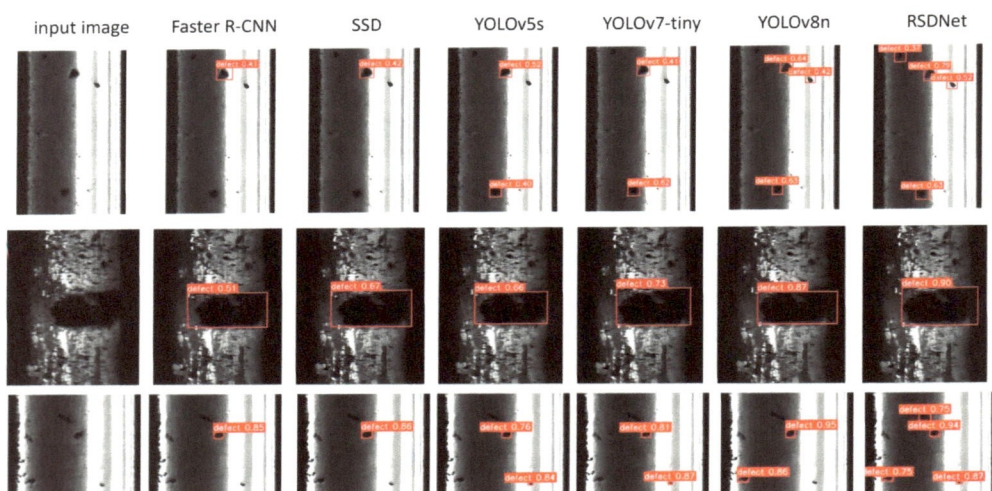

Figure 11. Comparison of the Detection Effect of Each Algorithm.

5. Conclusions

In this study, an advanced multi-scale rail surface defect detection model, RSDNet, is proposed to address the challenges of significant scale differences and numerous small-scale defects in the detection of rail surface defects. The essence of RSDNet's design lies in its excellent ability to capture multi-scale features, a feature that is crucial for accurately identifying small defects on the rail surface.

Based on YOLOv8n, RSDNet enhances the recognition of defects by designing the CDConv module, borrowing the BiFPN feature fusion approach, and introducing the EMA attention mechanism. Through these innovative improvements, RSDNet shows higher precision and recall in the task of track surface defect detection, and the algorithm proposed in this paper has higher accuracy, recall, and confidence in track surface defect recognition compared with other similar models. The multi-scale feature model performs well in terms of localization precision and detection precision, which can meet the practical requirements of real-time detection of track surface defects.

Looking ahead, the research team plans to collect more images of railroad track surface defects to enrich our dataset and refine the detailed descriptions of the defects. In addition, algorithms will be refined for images taken in strong-noise environments and blurred images that may arise from high-speed shooting to improve the generalization ability and robustness of the model. Efforts are made to apply the algorithms to embedded devices to detect defects on track surfaces in real time.

Author Contributions: Conceptualization, J.D.; methodology, R.Z. and J.D.; software, R.Z. and L.N.; formal analysis, R.G.; writing-review R.G. and Y.B. All authors have read and agreed to the published version of the manuscript.

Funding: This work was supported by the Natural Science Basic Research Program Project of Shaanxi Province, China (2024JC-YBMS-490).

Institutional Review Board Statement: Not applicable.

Informed Consent Statement: Not applicable.

Data Availability Statement: The rail surface dataset in this paper is a homemade dataset and is not disclosed due to its use in subsequent studies.

Conflicts of Interest: The authors declare no conflicts of interest.

References

1. Gong, W.; Akbar, M.F.; Jawad, G.N.; Mohamed, M.F.P.; Wahab, M.N.A. Nondestructive testing technologies for rail inspection: A review. *Coatings* **2022**, *12*, 1790. [CrossRef]
2. Oh, K.; Yoo, M.; Jin, N.; Ko, J.; Seo, J.; Joo, H.; Ko, M. A Review of Deep Learning Applications for Railway Safety. *Appl. Sci.* **2022**, *12*, 10572. [CrossRef]
3. Abbas, M.; Shafiee, M. Structural health monitoring (SHM) and determination of surface defects in large metallic structures using ultrasonic guided waves. *Sensors* **2018**, *18*, 3958. [CrossRef] [PubMed]
4. Park, J.W.; Lee, T.G.; Back, I.C.; Park, S.J.; Seo, J.M.; Choi, W.J.; Kwon, S.G. Rail surface defect detection and analysis using multi-channel eddy current method based algorithm for defect evaluation. *J. Nondestruct. Eval.* **2021**, *40*, 83. [CrossRef]
5. Alvarenga, T.A.; Carvalho, A.L.; Honorio, L.M.; Cerqueira, A.S.; Filho, L.M.; Nobrega, R.A. Detection and classification system for rail surface defects based on Eddy current. *Sensors* **2021**, *21*, 7937. [CrossRef]
6. Hao, S.; Gao, S.; Ma, X.; An, B.; He, T. Anchor-free infrared pedestrian detection based on cross-scale feature fusion and hierarchical attention mechanism. *Infrared Phys. Technol.* **2023**, *131*, 104660. [CrossRef]
7. Hao, S.; An, B.; Ma, X.; Sun, X.; He, T.; Sun, S. PKAMNet: A Transmission Line Insulator Parallel-Gap Fault Detection Network Based on Prior Knowledge Transfer and Attention Mechanism. *IEEE Trans. Power Deliv.* **2023**, *38*, 3387–3397. [CrossRef]
8. Shim, J.; Koo, J.; Park, Y. A Methodology of Condition Monitoring System Utilizing Supervised and Semi-Supervised Learning in Railway. *Sensors* **2023**, *23*, 9075. [CrossRef]
9. Fu, W.; He, Q.; Feng, Q.; Li, J.; Zheng, F.; Zhang, B. Recent advances in wayside railway wheel flat detection techniques: A review. *Sensors* **2023**, *23*, 3916. [CrossRef]
10. Xie, Q.; Tao, G.; He, B.; Wen, Z. Rail corrugation detection using one-dimensional convolution neural network and data-driven method. *Measurement* **2022**, *200*, 111624. [CrossRef]
11. Li, H.; Wang, Y.; Zeng, J.; Li, F.; Yang, Z.; Mei, G.; Ye, Y. Virtual point tracking method for online detection of relative wheel-rail displacement of railway vehicles. *Reliab. Eng. Syst. Saf.* **2024**, *246*, 110087. [CrossRef]
12. Ye, Y.; Zhu, B.; Huang, P.; Peng, B. OORNet: A deep learning model for on-board condition monitoring and fault diagnosis of out-of-round wheels of high-speed trains. *Measurement* **2022**, *199*, 111268. [CrossRef]
13. Xing, Z.; Zhang, Z.; Yao, X.; Qin, Y.; Jia, L. Rail wheel tread defect detection using improved YOLOv3. *Measurement* **2022**, *203*, 111959. [CrossRef]

14. Yang, H.; Wang, Y.; Hu, J.; He, J.; Yao, Z.; Bi, Q. Deep learning and machine vision-based inspection of rail surface defects. *IEEE Trans. Instrum. Meas.* **2021**, *71*, 5005714. [CrossRef]
15. Zhuang, L.; Qi, H.; Zhang, Z. The automatic rail surface multi-flaw identification based on a deep learning powered framework. *IEEE Trans. Intell. Transp. Syst.* **2021**, *23*, 12133–12143. [CrossRef]
16. Kou, L. A review of research on detection and evaluation of the rail surface defects. *Acta Polytech. Hung.* **2022**, *19*, 167. [CrossRef]
17. Gan, J.; Li, Q.; Wang, J.; Yu, H. A hierarchical extractor-based visual rail surface inspection system. *IEEE Sens. J.* **2017**, *17*, 7935–7944. [CrossRef]
18. Zhang, H.; Jin, X.; Wu, Q.J.; Wang, Y.; He, Z.; Yang, Y. Automatic visual detection system of railway surface defects with curvature filter and improved Gaussian mixture model. *IEEE Trans. Instrum. Meas.* **2018**, *67*, 1593–1608. [CrossRef]
19. Nieniewski, M. Morphological detection and extraction of rail surface defects. *IEEE Trans. Instrum. Meas.* **2020**, *69*, 6870–6879. [CrossRef]
20. Girshick, R.; Donahue, J.; Darrell, T.; Malik, J. Region-based convolutional networks for accurate object detection and segmentation. *IEEE Trans. Pattern Anal. Mach. Intell.* **2012**, *38*, 142–158. [CrossRef]
21. Girshick, R.; Donahue, J.; Darrell, T.; Malik, J. Rich feature hierarchies for accurate object detection and semantic segmentation. In Proceedings of the 2014 IEEE Conference on Computer Vision and Pattern Recognition, Columbus, OH, USA, 23–28 June 2014; pp. 580–587.
22. Ren, S.; He, K.; Girshick, R.; Sun, J. Faster r-cnn: Towards real-time object detection with region proposal networks. In Proceedings of the Advances in Neural Information Processing Systems 28: Annual Conference on Neural Information Processing Systems 2015, Montreal, QC, Canada, 7–12 December 2015.
23. He, K.; Gkioxari, G.; Dollár, P.; Girshick, R. Mask r-cnn. In Proceedings of the IEEE International Conference on Computer Vision (ICCV), Venice, Italy, 22–29 October 2017; pp. 2961–2969.
24. Cheng, Y.; HongGui, D.; YuXin, F. Effects of faster region-based convolutional neural network on the detection efficiency of rail defects under machine vision. In Proceedings of the 2020 IEEE 5th Information Technology and Mechatronics Engineering Conference (ITOEC), Chongqing, China, 12–14 June 2020; pp. 1377–1380.
25. Bai, T.; Yang, J.; Xu, G.; Yao, D. An optimized railway fastener detection method based on modified Faster R-CNN. *Measurement* **2021**, *182*, 109742. [CrossRef]
26. Wang, H.; Li, M.; Wan, Z. Rail surface defect detection based on improved Mask R-CNN. *Comput. Electr. Eng.* **2022**, *102*, 108269. [CrossRef]
27. Redmon, J.; Divvala, S.; Girshick, R.; Farhadi, A. You only look once: Unified, real-time object detection. In Proceedings of the 2016 IEEE Conference on Computer Vision and Pattern Recognition (CVPR), Las Vegas, NV, USA, 27–30 June 2016; pp. 779–788.
28. Redmon, J.; Farhadi, A. YOLO9000: Better, faster, stronger. In Proceedings of the 2017 IEEE Conference on Computer Vision and Pattern Recognition (CVPR), Honolulu, HI, USA, 21–26 July 2017; pp. 7263–7271.
29. Redmon, J.; Farhadi, A. Yolov3: An incremental improvement. *arXiv* **2018**, arXiv:1804.02767.
30. Bochkovskiy, A.; Wang, C.-Y.; Liao, H.-Y.M. Yolov4: Optimal speed and accuracy of object detection. *arXiv* **2020**, arXiv:2004.10934.
31. Li, C.; Li, L.; Jiang, H.; Weng, K.; Geng, Y.; Li, L.; Ke, Z.; Li, Q.; Cheng, M.; Nie, W. YOLOv6: A single-stage object detection framework for industrial applications. *arXiv* **2022**, arXiv:2209.02976.
32. Wang, C.-Y.; Bochkovskiy, A.; Liao, H.-Y.M. YOLOv7: Trainable bag-of-freebies sets new state-of-the-art for real-time object detectors. In Proceedings of the 2023 IEEE/CVF Conference on Computer Vision and Pattern Recognition (CVPR), Vancouver, BC, Canada, 17–24 June 2023; pp. 7464–7475.
33. Liu, W.; Anguelov, D.; Erhan, D.; Szegedy, C.; Reed, S.; Fu, C.-Y.; Berg, A.C. SSD: Single shot multibox detector. In Proceedings of the Computer Vision–ECCV 2016: 14th European Conference, Amsterdam, The Netherlands, 11–14 October 2016; Proceedings, Part I 14; Springer: Cham, Switzerland, 2016; pp. 21–37.
34. Wang, T.; Yang, F.; Tsui, K.-L. Real-time detection of railway track component via one-stage deep learning networks. *Sensors* **2020**, *20*, 4325. [CrossRef]
35. Wang, M.; Li, K.; Zhu, X.; Zhao, Y. Detection of surface defects on railway tracks based on deep learning. *IEEE Access* **2022**, *10*, 126451–126465. [CrossRef]
36. Zhang, C.; Xu, D.; Zhang, L.; Deng, W. Rail Surface Defect Detection Based on Image Enhancement and Improved YOLOX. *Electronics* **2023**, *12*, 2672. [CrossRef]
37. Wang, Y.; Zhang, K.; Wang, L.; Wu, L. An Improved YOLOv8 Algorithm for Rail Surface Defect Detection. *IEEE Access* **2024**, *12*, 44984–44997. [CrossRef]
38. Xin, F.; Jia, Q.; Yang, Y.; Pan, H.; Wang, Z. A high accuracy detection method for coal and gangue with S3DD-YOLOv8. *Int. J. Coal Prep. Util.* **2023**, 1–19. [CrossRef]
39. Yu, F.; Koltun, V. Multi-scale context aggregation by dilated convolutions. *arXiv* **2015**, arXiv:1511.07122.

40. Tan, M.; Pang, R.; Le, Q.V. Efficientdet: Scalable and efficient object detection. In Proceedings of the 2020 IEEE/CVF Conference on Computer Vision and Pattern Recognition (CVPR), Seattle, WA, USA, 13–19 June 2020; pp. 10781–10790.
41. Ouyang, D.; He, S.; Zhang, G.; Luo, M.; Guo, H.; Zhan, J.; Huang, Z. Efficient multi-scale attention module with cross-spatial learning. In Proceedings of the ICASSP 2023—2023 IEEE International Conference on Acoustics, Speech and Signal Processing (ICASSP), Rhodes Island, Greece, 4–10 June 2023; pp. 1–5.

Disclaimer/Publisher's Note: The statements, opinions and data contained in all publications are solely those of the individual author(s) and contributor(s) and not of MDPI and/or the editor(s). MDPI and/or the editor(s) disclaim responsibility for any injury to people or property resulting from any ideas, methods, instructions or products referred to in the content.

Article

Real-Time Trajectory Smoothing and Obstacle Avoidance: A Method Based on Virtual Force Guidance

Yongbin Su, Chenying Lin and Tundong Liu *

Pen-Tung Sah Institute of Micro-Nano Science and Technology, Xiamen University, Xiamen 361104, China; suyongbin@stu.xmu.edu.cn (Y.S.); linchenying@stu.xmu.edu.cn (C.L.)
* Correspondence: ltd@xmu.edu.cn; Tel.: +86-130-3089-4446

Abstract: In dynamic environments, real-time trajectory planners are required to generate smooth trajectories. However, trajectory planners based on real-time sampling often produce jerky trajectories that necessitate post-processing steps for smoothing. Existing local smoothing methods may result in trajectories that collide with obstacles due to the lack of a direct connection between the smoothing process and trajectory optimization. To address this limitation, this paper proposes a novel trajectory-smoothing method that considers obstacle constraints in real time. By introducing virtual attractive forces from original trajectory points and virtual repulsive forces from obstacles, the resultant force guides the generation of smooth trajectories. This approach enables parallel execution with the trajectory-planning process and requires low computational overhead. Experimental validation in different scenarios demonstrates that the proposed method not only achieves real-time trajectory smoothing but also effectively avoids obstacles.

Keywords: trajectory planning; real-time trajectory smoothing; real-time obstacle avoidance; virtual force guidance

Citation: Su, Y.; Lin, C.; Liu, T. Real-Time Trajectory Smoothing and Obstacle Avoidance: A Method Based on Virtual Force Guidance. *Sensors* 2024, 24, 3935. https://doi.org/10.3390/s24123935

Academic Editors: Chunhui Zhao and Shuai Hao

Received: 22 May 2024
Revised: 7 June 2024
Accepted: 14 June 2024
Published: 18 June 2024

Copyright: © 2024 by the authors. Licensee MDPI, Basel, Switzerland. This article is an open access article distributed under the terms and conditions of the Creative Commons Attribution (CC BY) license (https://creativecommons.org/licenses/by/4.0/).

1. Introduction

Trajectory planning entails determining a path between two points within a specified region while avoiding collisions with environmental objects [1]. It is pivotal for various unmanned platforms, including industrial robots, mobile robots, and autonomous drones. The smoothness of trajectories is a critical performance metric in trajectory planning, as jerky trajectories can lead to frequent transitions between "stop", "rotate", and "restart" motion states, resulting in slip and over-responsive behavior during high-speed movements [2,3]. In dynamic environments, unpredictable changes occur frequently, necessitating online adjustments or recalibrations of paths to ensure safe navigation around newly detected nearby objects [4]. Hence, there is a critical need for efficient online trajectory-smoothing and obstacle-avoidance methods to meet the post-processing requirements of real-time planning based on sampling.

To address trajectory planning in dynamic environments, unmanned platforms need to identify potential paths based on obstacle information at each time step and determine the next foothold until reaching the target position [5]. While real-time trajectory planning allows adaptive responses to sudden environmental changes, the lack of smoothness makes them inefficient and prone to system instability [6]. Smooth trajectories are desirable in trajectory planning for practical unmanned platforms as they appear more predictable and satisfactory to humans [7]. Existing trajectory-smoothing methods can be categorized into global and local smoothing approaches [8]. Global smoothing fits all trajectory points using a parameterized curve (e.g., NURBS [9], polynomial splines [10], or B-splines [8]) after obtaining the complete trajectory, which is impractical for real-time trajectory-planning scenarios. Local smoothing methods aim to replace sharp corners within two adjacent line segments with smooth blending curves. Various local smoothing algorithms have been proposed, incorporating different curve representations to ensure smooth transitions at trajectory corners [11–16]. However, these methods may produce paths that collide with obstacles (as indicated in the situation shown in Figure 1) since the smoothing process lacks a direct connection with trajectory optimization [15].

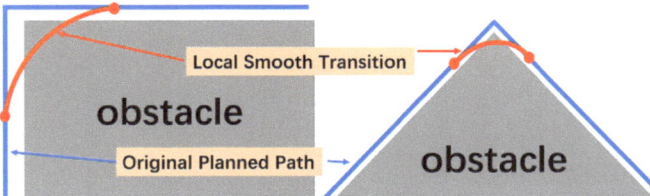

Figure 1. The collision situations between the trajectories generated by the local smoothing algorithm and obstacles.

To address the real-time smoothing and obstacle-avoidance problem, many researchers have studied real-time trajectory smoothers based on shortcut techniques. The idea of shortcutting a smoothed path was initially proposed by Geraerts and Overmars [17]. Hauser et al. [18] extended this approach by introducing parabolic trajectory representations to consider velocity and acceleration bounds. Subsequently, Ran et al. [19] further expanded this parabolic smoothing algorithm to incorporate jerk constraints. Furthermore, Pan et al. [20] introduced spline-based trajectory representation and its associated shortcut algorithm. Although all of these methods successfully compute smooth paths for piecewise linear trajectories, their computation speed is slow due to numerous collision checks. To optimize computation speed, Shohei et al. [6] employed neural networks to estimate gaps and collisions between a robot and voxels in a parallel manner, combined with shortcut techniques to construct a real-time trajectory smoother, achieving a cycle time of 300 ms. However, this approach requires high hardware requirements and extensive dataset collection for training the network model, limiting its practicality.

Here, we propose a novel trajectory-smoothing framework to provide real-time trajectory smoothing for trajectory planners based on real-time sampling, achieving obstacle avoidance and trajectory smoothing in dynamic environments. This method constructs a "follower" to track the trajectory points generated by the trajectory planner, where points ahead of the "follower" exert attractive forces, pulling it to follow the original trajectory, while points behind the "follower" exert repulsive forces, pushing it forward. When obstacles appear, a repulsive force is applied to the "follower" to encourage it to move away from the obstacles. The resultant force from all forces determines the forward direction of the "follower". By appropriately setting the follower's velocity and introducing a lag effect at corners, trajectory smoothing is achieved. The position of the "follower" is recorded at each time step, constituting the final smoothed trajectory.

2. Proposed Method

2.1. The Overall Framework of the Proposed Method

The overall framework of the method proposed in this paper is illustrated in Figure 2. It explains the effectiveness, principles, and processes of this method. The performance of the proposed real-time trajectory smoothing method based on virtual force guidance is shown in the upper left part of Figure 2. The trajectory planner in the figure adopts a real-time sampling-based trajectory-planning method, which calculates the position of the next point at each subsequent time step until reaching the goal point while avoiding obstacles. The trajectory smoother, employing the method proposed in this paper, operates concurrently with the trajectory planner. Whenever the planner generates a new sampling point, the smoother incorporates it and calculates the next position using the smoothing algorithm.

The smoother first defines a "follower" that follows the trajectory points generated by the trajectory planner from time t_0 according to certain rules. Specifically, as shown in the upper right part of Figure 2, the point generated by the planner at a certain moment and the points generated in the previous three time steps (labeled A, B, C, and D, respectively), along with the obstacles, jointly influence the output of the smoother. Points A and B exert attractive forces on the "follower", while the other points exert repulsive forces. The resultant force guides the "follower" to the position at the subsequent time step, continuing until reaching the goal point. The overall process of this method is shown in the lower part of Figure 2.

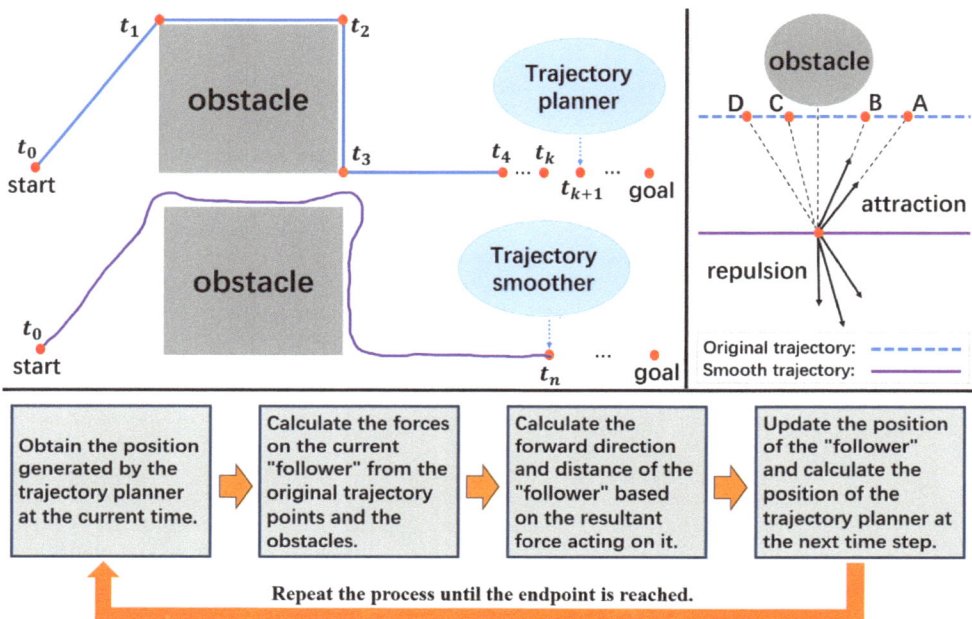

Figure 2. The overall framework of the proposed method.

2.2. Analysis of Obstacle-Free Scenarios

A concept known as the reference point is introduced into the original trajectory (generated by the trajectory planner), akin to the role of point B in Figure 2; this serves as the primary attractor for the "follower". As shown in Figure 3, assuming that the red point represents the "follower" (the current position of the unmanned platform, i.e., the goal point of the smoothed trajectory), denoted as P, and the blue point represents the reference point of the original trajectory, denoted as x_k, with its subsequent point being denoted as x_{k+1} and its preceding points being denoted as x_{k-1} and x_{k-2}. The forces that they exert on point P are denoted as follows.

$$\vec{F}_k = k_{at1} \cdot (x_k - P) \tag{1}$$

$$\vec{F}_{k+1} = k_{at2} \cdot (x_{k+1} - P) \tag{2}$$

$$\vec{F}_{k-1} = k_{re1} \cdot (x_{k-1} - P) \tag{3}$$

$$\vec{F}_{k-2} = k_{re2} \cdot (x_{k-2} - P) \tag{4}$$

where k_{at1}, k_{at2}, k_{re1}, and k_{re2} represent the coefficients of attractive and repulsive forces; the resultant force F_{ori} experienced by the "follower" can be calculated using the following formula.

$$\vec{F_{ori}} = \vec{F_k} + \vec{F_{k+1}} + \vec{F_{k-1}} + \vec{F_{k-2}} \tag{5}$$

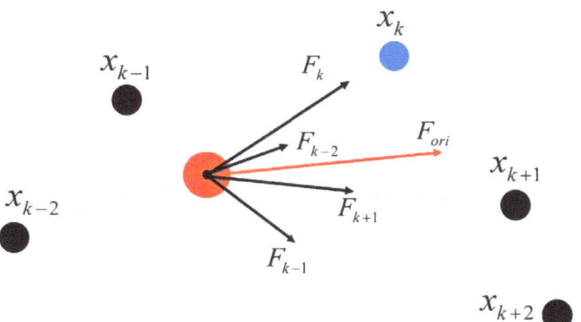

Figure 3. Analytical diagram of the virtual forces exerted on the "follower" by points in the original trajectory.

The direction of vector $\vec{F_{ori}}$ serves as the direction of motion for P, and its magnitude determines the motion step length. P will move forward to reach the position of the red point in Figure 4a. If, at this point, the reference point remains unchanged, the process continues with force analysis controlling the forward movement of P, as described above. If the reference point updates to the next point (i.e., changing from the original x_k to x_{k+1}), then P undergoes force analysis and motion under the influence of the new reference point, as shown in Figure 4b. The specific strategy for updating the reference point will be discussed in Section 2.4.

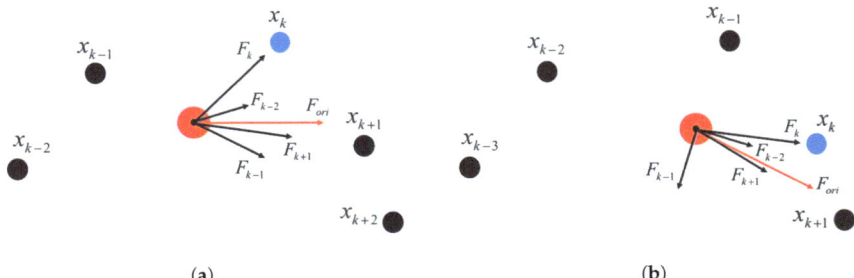

(a) (b)

Figure 4. Analysis of the forces acting on the "follower" and its motion at different reference points. Panel (**a**) depicts the force analysis diagram when the reference point remains unchanged. Panel (**b**) illustrates the force analysis diagram when the reference point is updated to the next trajectory point.

2.3. Analysis in the Presence of Obstacles

For irregularly shaped obstacles, the current mainstream approach is to replace them with a circumcircle [21], as circular obstacle boundaries are more conducive to generating smooth trajectories. However, if only a circumcircle of obstacles is constructed to replace the original shape, it will encroach on a significant amount of free space, resulting in overly lengthy generated trajectories. As shown in Figure 5a, the circumcircles of two irregularly shaped obstacles block the gap between them, forcing the planned trajectory to detour around them (solid lines in the figure), rather than passing through the gap (dashed lines in the figure), which is undesirable.

To address this issue, we adopt a more efficient approach by rounding the corners of irregular obstacles with circles of radius r, as shown in Figure 5b. This not only makes the obstacle surface smoother, facilitating subsequent algorithm execution, but also minimizes the space occupied, thus preserving the optimal solution as much as possible. When the position P of the "follower" is within the influence range of obstacles, it is necessary to introduce the effect of obstacle repulsion to encourage the points generated by the trajectory smoother to move away from obstacles. The black arrows in the figure represent the direction of the repulsive force exerted by the obstacles on the surroundings, which are perpendicular to the contours of the obstacles. When the "follower" approaches the range of action of the obstacles, it will be affected by the repulsive force of the obstacles. The point where the repulsive force is generated is the point on the contour of the obstacle that is closest to it. If there is more than one such point, then the repulsive force is the resultant force of all points acting on it. As shown by the blue dashed circle in the figure, suppose that the "follower" moves to the position where the red dot is located. There are two points on the contour of the obstacle closest to it, so it will be affected by two forces, F_1 and F_2. The final repulsive force exerted by the obstacle on it is the resultant force of these two forces, denoted as F_{obs}.

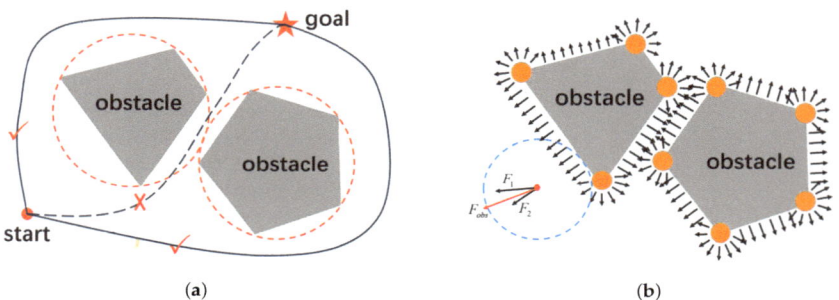

Figure 5. Different approaches to handling irregular obstacle corners. Panel (**a**) depicts covering the original obstacle with a circumscribed circle, allowing the trajectory to pass only around the obstacle edges and making it unable to penetrate through the center. Panel (**b**) illustrates placing a circular obstacle at each corner of the obstacle and analyzing the force acting on objects within the obstacle's influence range.

For the sake of simplicity in modeling and description, in the subsequent content, we still assume that all obstacles are circular. Discussing the general case, when the planned trajectory passes through obstacles, as shown in Figure 6a, in addition to the forces exerted on point P by the four points of the original trajectory, the obstacles also exert forces on it. The resultant force of these components guides the extension direction of the smooth trajectory.

Similarly to the artificial potential field (APF) method, obstacles typically have an influence range ρ, and only objects within this range are affected by the repulsive force of the obstacles. However, there is a scenario where the "follower" enters the influence range of the obstacles, but its direction of motion has never pointed toward the obstacles and just passes by. If such instances are influenced by the obstacles, it can lead to unnecessary deviations in the smooth trajectory. Therefore, this study determines whether the obstacles will affect the "follower" by examining the number of intersection points between the direction of motion of the "follower" and the contour of the obstacles.

Suppose that the influence radius of the obstacle is ρ, as shown in Figure 6b. At time k, the position of the "follower" is y_k, and its position at the previous time is y_{k-1}. Connecting these two points into a straight line, it is assumed that they intersect with the obstacle at points M and N and intersect with the boundary of the influence range at point O. If points M or N exist, then when analyzing the force acting on y_k, the repulsive force will be considered. We define the distance from y_k to the nearest intersection point O as d_1 and the distance from O to M as d_2. Then, the repulsive force acting on y_k is defined as

$$|\vec{F_{obs}}| = d_1/d_2 * F_{max} \tag{6}$$

where F_{max} represents the maximum value of the repulsive force, that is, the value when y_k comes into contact with the surface of the obstacle. It is not fixed but has a certain relationship with F_{ori}, $F_{max} = k * F_{ori}$, to prevent sudden changes in the final resultant force when approaching the obstacle. The direction of $\vec{F_{obs}}$ is from the center of the obstacle to y_k. If there are no intersection points, then the repulsive force at this time will be zero.

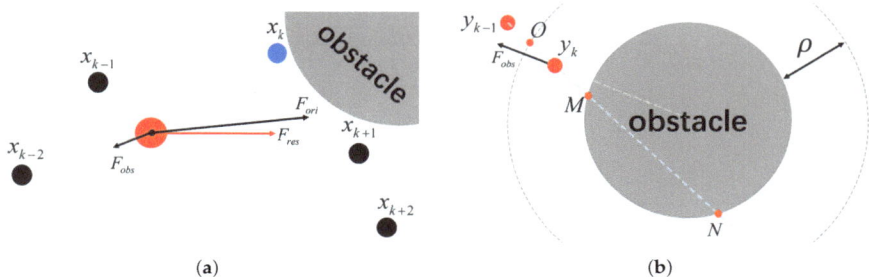

(a) (b)

Figure 6. Force analysis in the presence of obstacles. Panel (**a**) depicts the overall force acting on the "follower" when obstacles are introduced. Panel (**b**) separately analyzes the calculation of the magnitude and direction of the force exerted by obstacles on the follower.

2.4. Reference Point Update Strategy

In the absence of obstacles, relying solely on the virtual force guidance provided by the original trajectory, the "follower" may closely track the original path, potentially overlapping with it, thus failing to achieve the desired smoothing effect. To address this issue, we introduce a maximum tracking error parameter *disErr* to ensure that the original trajectory slightly leads the smoothed trajectory. Consequently, when the original trajectory encounters corners, the smoothed trajectory, due to its inherent lag, can smoothly transition around them. If the distance between the reference point of the original trajectory and the "follower" exceeds a certain threshold, the update frequency of the reference point is reduced by a factor of v_0 until their distance meets the tracking error requirement. This approach not only ensures smooth transitions around corners but also prevents excessive distortion of the smoothed trajectory.

When approaching obstacles, the update of the reference point needs to be redesigned. If the sampling points output by the trajectory planner are too close to obstacles and are obstructed by circles covering the corner points, the original trajectory may penetrate through the obstacles. In such cases, the smoothed trajectory needs to navigate around the obstacles. This may result in a longer path compared to the original trajectory, and using the update strategy from the obstacle-free scenario may lead to oscillations in the output points of the trajectory smoother. Our strategy is to adjust the update frequency of the reference point based on the distance between the reference point and the center of the obstacle, categorized into three levels $\{v_1, v_2, v_3\}$, as illustrated in Figure 7. Specifically, the update frequency is accelerated when entering the influence range of obstacles, reduced when close to the center of the obstacles, and restored to the baseline update frequency when moving away from obstacles.

In addition to setting the maximum tracking error and the update frequency of the reference point, it is necessary to define a maximum step size parameter *stepMax* to adjust the feed rate of the "follower" to prevent abrupt changes in trajectory caused by excessively large step sizes. If the calculated feed rate exceeds this limit, the step size in each direction is proportionally reduced. Furthermore, since the forward direction and step size of the "follower" are guided by the resultant force, situations where the resultant force equals zero may occur, which are known as local optima in the APF method. To avoid this, a small force $\vec{F_1}$ perpendicular to $\vec{F_{ori}}$ is introduced into the force calculation. Algorithm 1 outlines this process.

Algorithm 1: Real-Time Smoothing Algorithm

Input: Points computed by the trajectory planner at the current time and trajectory points obtained from the previous three time steps: $\{x_{k+1}, x_k, x_{k-1}, x_{k-2}\}$; the position of the follower: y_f; maximum step size: $stepMax$; maximum tracking error: $disErr$; parameters related to the reference point update frequency: $\{v_0, v_1, v_2, v_3\}$.

Output: The set of smoothed trajectory points $\{y_k\}$ at the current time.

Parameter initialization: count=0

while *true* **do**
    count = count + 1
    if $|x_{k+1} - x_{goal}| > 0.1$ **then**
        (1) Calculate the resultant force acting on the "follower" based on Equations (5) and (6): $\vec{F_{res}} = \vec{F_{ori}} + \vec{F_{obs}} + \vec{F_1}$
        (2) **if** $|\vec{F_{res}}| > stepMax$ **then**
            $step = \frac{stepMax}{|\vec{F_{res}}|} \vec{F_{res}}$
        end
        else
            $step = \vec{F_{res}}$
        end
        (3) $y_f = y_f + step$, $\{y_k\} \leftarrow y_f$
        (4) **if** $y_f \in S_0$ **then**
            if $|y_f - x_k| < disErr$ **then**
                break while
            end
        else
            if *count* mod $\frac{1}{v_0} = 0$ **then**
                break while
            end
        end
    end
    else if $y_f \in S_1$ **then**
        Waiting for planner to compute the next v_1-th point: $x_{k+1} = x_{k+1+v_1}$;
        break while
    end
    else if $y_f \in S_2$ **then**
        if *count* mod $\frac{1}{v_2} = 0$ **then**
            break while
        end
    end
    else if $y_f \in S_3$ **then**
        if *count* mod $\frac{1}{v_3} = 0$ **then**
            break while
        end
    end
end

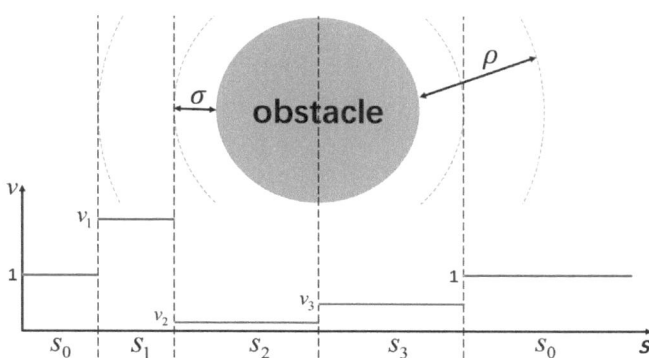

Figure 7. The update frequency of the reference points from the original trajectory within the range of influence of obstacles.

3. Experimental Results

3.1. Experimental Setup and Evaluation Metrics

We conducted simulation experiments in a two-dimensional plane. The starting point is represented by blue dots, the destination point is represented by red dots, and obstacles are represented by black-filled circles with shaded regions indicating their influence range. Since we focus solely on the smoothing process of known trajectories rather than trajectory planning, we preset the trajectory path and simulate the real-time planning process, where yellow dots represent points generated at each time step. Our trajectory smoother processes these yellow points in real time, smoothing them at corners; these are represented by blue dots in the experiment. We only present the final processed trajectory, while the specific process can be observed by running our open-source code (https://github.com/syb-xmu/Real-time-Smoothing.git).

In the absence of obstacles, local smoothing methods such as Bezier curves effectively smooth corners. Therefore, in this scenario, we focus on analyzing the influence of our algorithm's parameters on the experimental results. In the presence of obstacles, we compare our method with RRT–Bezier-curve-based and APF-based methods. However, since these methods require the destination to be known beforehand, we conduct obstacle-avoidance processing only after the trajectory planner passes through obstacles. We define three quantitative evaluation metrics to assess our algorithm. The curvature measures trajectory smoothness and is calculated as described in [22]. The Fréchet distance [23] evaluates the similarity between the original and smoothed trajectories. The collision probability assesses the algorithm's obstacle-avoidance performance and is calculated as the ratio of collision instances to total experimental trials.

3.2. Case Study 1: No-Obstacle Scenario

In this section, we explore the impacts of two critical parameters, namely, the maximum step size and the maximum tracking error, on the algorithm's performance. We assume that the trajectory planner generates a trajectory resembling a sawtooth wave, and our algorithm performs real-time smoothing around corners. First, we investigate the influence of different maximum tracking errors, setting their values to 0.5, 1.0, and 2.0, while keeping other parameters constant: $k_{at1} = 1$, $k_{at2} = 0.5$, $k_{re1} = k_{re2} = 0.01$, $k = 1.1$, $v_0 = \frac{1}{3}$, $v_1 = 2$, $v_2 = \frac{1}{6}$, $v_3 = \frac{1}{2}$, and $stepMax = 0.1$. The obtained results and corresponding curvature are shown in Figure 8.

Since the number of points in the smoothed trajectory varies for different parameters, the relative positions of the smoothed trajectory and the original trajectory in the curvature plot also differ. From the results, it can be observed that a larger maximum tracking error leads to smoother trajectories but also introduces greater distortion. Specific quantitative metrics are shown in Table 1, indicating an inverse relationship between trajectory smoothness and distortion. Hence, suitable parameters can be chosen to balance these factors according to practical requirements.

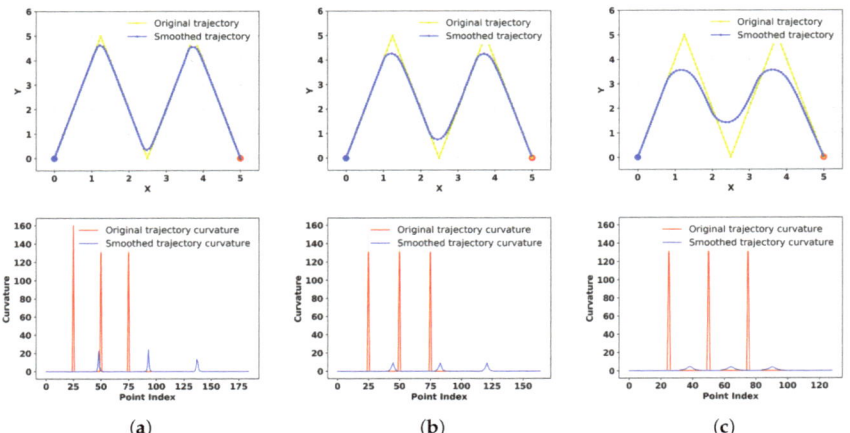

Figure 8. The smoothed trajectory directions and corresponding curvature values with different maximum tracking errors (**a**–**c**).

Table 1. The trajectory similarity and smoothness with different maximum tracking errors.

Maximum Tracking Error	Fréchet Distance	Max Curvature
0.5	0.45	24.45
1.0	0.76	9.49
2.0	1.45	4.47

In our algorithm, the maximum step size is also a crucial parameter affecting trajectory smoothness. For the same corner, if the maximum step size is too large, smooth transitions cannot be achieved; if it is too small, the trajectory may not keep up, leading to distortion. We further investigate the influence of different maximum step sizes, setting their values to 0.12, 0.15, and 0.18 while setting $disErr = 1$, with the other parameters remaining unchanged. The obtained results and corresponding curvature are depicted in Figure 9.

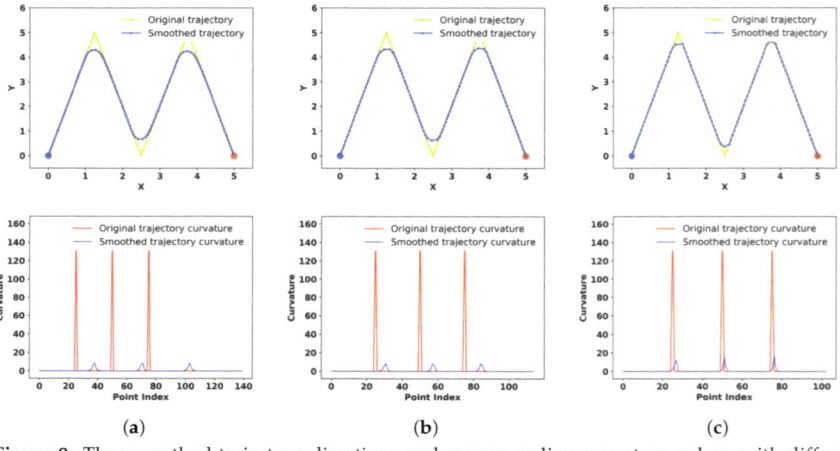

Figure 9. The smoothed trajectory directions and corresponding curvature values with different maximum step sizes (**a**–**c**).

From the results, it can be observed that larger maximum step sizes result in more pronounced corners in the trajectory, indicating poorer smoothness. Specific quantitative metrics are shown in Table 2, where an increase in the maximum step size leads to a closer match between the extension speed of the smoothed trajectory and the original trajectory, without any lag effect. Consequently, although the trajectories exhibit high similarity, the smoothness is compromised.

Table 2. The trajectory similarity and smoothness with different maximum step sizes.

Maximum Step Size	Fréchet Distance	Max Curvature
0.12	0.75	9.06
0.15	0.68	8.33
0.18	0.49	17.49

3.3. Case Study 2: Single-Obstacle Straight Trajectory Scenario

This experiment investigates the performance of our method in a simple scenario where the trajectory passes through an obstacle in a straight line. We compare our method with traditional RRT algorithms, RRT algorithms with Bezier curves, and traditional APF methods to validate the superior obstacle avoidance and smoothing performance of our approach. Traditional RRT algorithms, due to their randomness, yield different results each time and may generate trajectories with peculiar shapes, as shown in Figure 10a. Incorporating Bezier curves can achieve smoothing effects, but collisions with obstacles may occur, as illustrated in Figure 10b, because the post-smoothing process is not directly connected to the trajectory-planning process and does not consider obstacle constraints. We ran the RRT algorithm 20 times and selected the best results for comparison with our method, as depicted in Figure 10c. Additionally, we calculated the collision probability by counting the number of collisions that occurred after smoothing with Bezier curves.

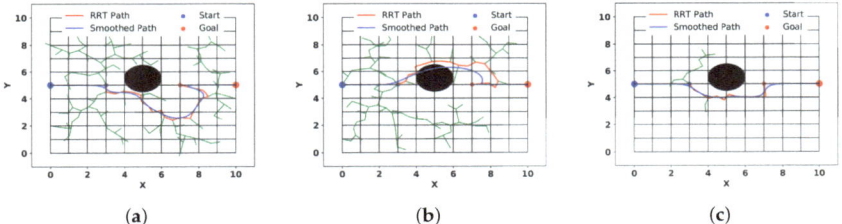

Figure 10. The trajectories obtained from the smoothing methods based on the RRT algorithm and Bezier curves.

The trajectory generated by the traditional APF method tends to deviate significantly from obstacles due to the lack of constraints, resulting in a large deviation from the ideal trajectory. To ensure fairness, we also use Equation (6) to replace the traditional APF method's obstacle repulsion calculation. Additionally, to adapt the traditional APF method to our scenario, we set the entry point into the obstacle's influence area as the starting point and designate the exit point as the goal. Between these two points, we employ the APF method for the transition. The trajectories obtained with traditional APF methods and our algorithm, as shown in Figure 11a,b, reveal that the trajectories generated by traditional APF methods exhibit a directional discontinuity when approaching the lowest point of the obstacle, while our method smoothly circumvents the obstacle. The curvature plots obtained from these four methods, as shown in Figure 11c, demonstrate that our method has the lowest maximum curvature.

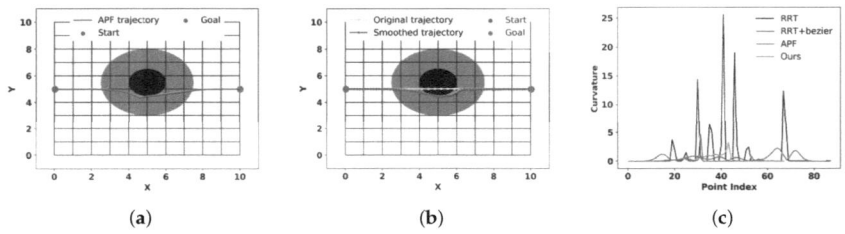

Figure 11. (**a**) Trajectory obtained from artificial potential field (APF). (**b**) Trajectory obtained from the proposed method. (**c**) Curvatures of trajectories obtained from different methods.

Comparing the results in Figure 10c with those in Figure 11a,b, we observe that among these four methods, the trajectories obtained with the RRT-based method can easily avoid obstacles, but they appear stiff, and collisions may occur after smoothing with Bezier curves. The traditional APF method ensures collision-free trajectories but exhibits directional discontinuities at specific locations, resulting in non-smooth trajectories. Table 3 presents a quantitative comparison of the results of different methods. In terms of trajectory similarity, both the traditional APF method and our method demonstrate superior performance. However, our method has a lower maximum curvature, indicating better smoothness. Additionally, the traditional RRT- and Bezier-curve-based methods experienced collisions in 6 out of the 20 experiments, indicating a significant risk despite their superior smoothness. In summary, our method outperforms others in terms of trajectory similarity and smoothness, and most importantly, it achieves real-time smoothing effects, while other methods require separate obstacle avoidance and smoothing processes after the trajectory planner completely bypasses the obstacles.

Table 3. Trajectory similarity, smoothness, and collision probability with different methods.

Method	Fréchet Distance	Max Curvature	Collision Probability
RRT	1.19	25.56	0%
RRT+Bezier	1.03	2.25	30%
APF	0.64	4.63	0%
Ours	0.56	3.22	0%

3.4. Case Study 3: Single-Obstacle Corner Trajectory Scenario

To minimize the collision space caused by obstacles and to eliminate some of the sharp corners of the original obstacles, we use circular obstacles to cover the corners of the original obstacles (as described in Section 2.3). However, this inevitably obstructs the trajectories planned by the trajectory planner. Therefore, in this experiment, we explore the obstacle avoidance and smoothing performance of our algorithm with different corner angles. As the previous section's experiment showed, the trajectory smoothness of the traditional RRT algorithm is poor, and adding Bezier curves may lead to collisions with a relatively high probability. Therefore, in this section, we only compare our method with the APF method.

Corner angles of 45 degrees, 90 degrees, and 135 degrees are set, and the results are shown in Figure 12. It can be seen that the trajectories generated by our method smoothly navigate around the obstacles, and the trajectory directions are reasonable. However, the traditional APF method cannot consider a trajectory inside the obstacle, so it directly selects the original destination point (i.e., the point where the original trajectory exits the obstacle) as the attractor point, resulting in an unreasonable trajectory. In contrast, our method dynamically follows the virtual target points of the original trajectory, so it can reproduce the trajectory's path as much as possible, making the smoothed trajectory more reasonable and of higher quality.

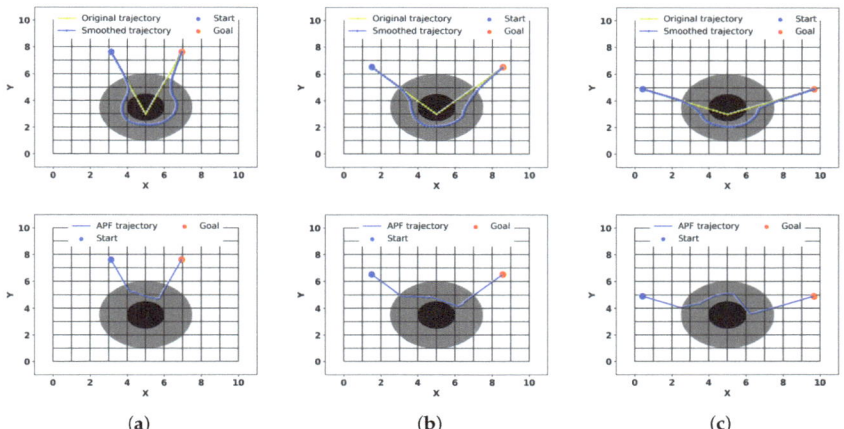

(a)   (b)   (c)

Figure 12. Smooth trajectories obtained with the proposed method and the APF method at different angles (**a–c**).

3.5. Case Study 4: Multi-Obstacle Complex Trajectory Scenario

Our method also applies to situations where unexpected obstacles suddenly appear in the original trajectory, which is common in the field of learning from demonstrations. In this experiment, we assume that the original trajectory follows a sinusoidal function shape, and at some point, two obstacles suddenly appear. Using our method, the trajectory can smoothly navigate around these obstacles. Figure 13 illustrates two scenarios with obstacles at different positions. In Figure 13a, the obstacle on the left does not obstruct the original trajectory directly, but the original trajectory enters its influence range. However, since we only consider repulsive forces when the direction of motion intersects with the corner of the obstacle, as described in Section 2.3, entering the influence area of the obstacle does not change the trajectory's direction.

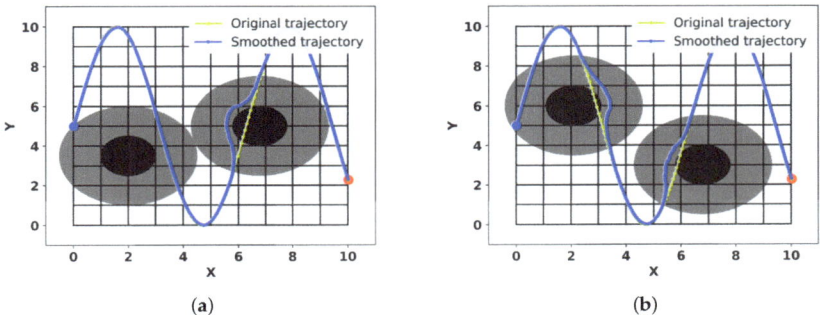

(a)   (b)

Figure 13. Smooth obstacle-avoidance trajectories generated by the proposed method at different obstacle positions (**a**,**b**).

4. Discussion and Conclusions

This paper proposes a real-time trajectory-smoothing framework based on virtual force guidance that can be executed in parallel with real-time trajectory-planning algorithms. By constructing a virtual force model between the original trajectory and obstacles, it guides the real-time generation of smooth trajectories. In the absence of obstacles, the proposed method can smoothly navigate the corners of the original trajectory. When obstacles appear, comparisons with RRT and APF algorithms reveal that our method not only smoothly navigates obstacles but also maintains the direction of the original trajectory well, successfully avoiding obstacles and, thereby, validating the effectiveness and superiority of our method.

Our method is not only applicable to real-time trajectory smoothing but also to real-time smoothing of sampled signals to remove signal spikes and high-frequency noise. Additionally, our method is suitable for obstacle avoidance in pre-planned trajectories, especially in the field of demonstration learning. If pre-planned trajectories are established and obstacles suddenly appear during execution, our method can effectively avoid them. In the absence of obstacles, simply reducing the maximum tracking error can essentially replicate the original trajectory. Furthermore, by appropriately setting the maximum step size, the acceleration and deceleration processes can be seamlessly integrated into the proposed method without the need for subsequent interpolation processing, making it highly suitable for practical applications in unmanned platforms.

Author Contributions: Conceptualization, Y.S. and T.L.; methodology, Y.S. and C.L.; software, Y.S.; validation, Y.S.; formal analysis, Y.S. and T.L.; investigation, Y.S. and C.L.; resources, Y.S.; data curation, Y.S.; writing—original draft preparation, Y.S.; writing—review and editing, T.L.; visualization, C.L.; supervision, C.L.; project administration, T.L. All authors have read and agreed to the published version of the manuscript.

Funding: This research received no external funding.

Institutional Review Board Statement: Not applicable.

Informed Consent Statement: Not applicable.

Data Availability Statement: Data are contained within the article. We have open-sourced our code on GitHub at https://github.com/syb-xmu/Real-time-Smoothing.git.

Conflicts of Interest: The authors declare no conflicts of interest.

References

1. Lu, E.; Tian, Z.; Xu, L.; Ma, Z.; Luo, C. Observer-based robust cooperative formation tracking control for multiple combine harvesters. *Nonlinear Dyn.* **2023**, *111*, 15109–15125. [CrossRef]
2. Berglund, T.; Brodnik, A.; Jonsson, H.; Staffanson, M.; Soderkvist, I. Planning smooth and obstacle-avoiding B-spline paths for autonomous mining vehicles. *IEEE Trans. Autom. Sci. Eng.* **2009**, *7*, 167–172. [CrossRef]
3. Elhoseny, M.; Tharwat, A.; Hassanien, A.E. Bezier curve based path planning in a dynamic field using modified genetic algorithm. *J. Comput. Sci.* **2018**, *25*, 339–350. [CrossRef]
4. Sung, I.; Choi, B.; Nielsen, P. On the training of a neural network for online path planning with offline path planning algorithms. *Int. J. Inf. Manag.* **2021**, *57*, 102142. [CrossRef]
5. Schmid, L.; Pantic, M.; Khanna, R.; Ott, L.; Siegwart, R.; Nieto, J. An efficient sampling-based method for online informative path planning in unknown environments. *IEEE Robot. Autom. Lett.* **2020**, *5*, 1500–1507. [CrossRef]
6. Fujii, S.; Pham, Q.C. Realtime Trajectory Smoothing with Neural Nets. In Proceedings of the 2022 International Conference on Robotics and Automation (ICRA), Xiamen, China, 23–27 May 2022; pp. 7248–7254.
7. Pham, Q.-C.; Nakamura, Y. A new trajectory deformation algorithm based on affine transformations. *IEEE Trans. Robot.* **2015**, *31*, 1054–1063. [CrossRef]
8. Xiao, Q.-B.; Wan, M.; Qin, X.-B.; Liu, Y.; Zhang, W.-H. Real-time smoothing of G01 commands for five-axis machining by constructing an entire spline with the bounded smoothing error. *Mech. Mach. Theory* **2021**, *161*, 104307. [CrossRef]
9. Yeh, S.-S.; Su, H.-C. Implementation of online NURBS curve fitting process on CNC machines. *Int. J. Adv. Manuf. Technol.* **2009**, *40*, 531–540. [CrossRef]
10. Yuen, A.; Zhang, K.; Altintas, Y. Smooth trajectory generation for five-axis machine tools. *Int. J. Mach. Tools Manuf.* **2013**, *71*, 11–19. [CrossRef]
11. Fan, W.; Ji, J.; Wu, P.; Wu, D.; Chen, H. Modeling and simulation of trajectory smoothing and feedrate scheduling for vibration-damping CNC machining. *Simul. Model. Pract. Theory* **2020**, *99*, 102028. [CrossRef]
12. Han, J.; Jiang, Y.; Tian, X.; Chen, F.; Lu, C.; Xia, L. A local smoothing interpolation method for short line segments to realize continuous motion of tool axis acceleration. *Int. J. Adv. Manuf. Technol.* **2018**, *95*, 1729–1742. [CrossRef]
13. Hu, Q.; Chen, Y.; Jin, X.; Yang, J. A real-time C3 continuous local corner smoothing and interpolation algorithm for CNC machine tools. *J. Manuf. Sci. Eng.* **2019**, *141*, 041004. [CrossRef]
14. Jiang, X.; Hu, Y.; Huo, G.; Su, C.; Wang, B.; Li, H.; Shen, L.; Zheng, Z. Asymmetrical Pythagorean-hodograph spline-based C4 continuous local corner smoothing method with jerk-continuous feedrate scheduling along linear toolpath. *Int. J. Adv. Manuf. Technol.* **2022**, *121*, 5731–5754. [CrossRef]

15. Song, B.; Wang, Z.; Zou, L. An improved PSO algorithm for smooth path planning of mobile robots using continuous high-degree Bezier curve. *Appl. Soft Comput.* **2021**, *100*, 106960. [CrossRef]
16. Wang, H.; Wu, J.; Liu, C.; Xiong, Z. A real-time interpolation strategy for transition tool path with C2 and G2 continuity. *Int. J. Adv. Manuf. Technol.* **2018**, *98*, 905–918. [CrossRef]
17. Geraerts, R.; Overmars, M.H. Creating high-quality paths for motion planning. *Int. J. Robot. Res.* **2007**, *26*, 845–863. [CrossRef]
18. Hauser, K.; Ng-Thow-Hing, V. Fast smoothing of manipulator trajectories using optimal bounded-acceleration shortcuts. In Proceedings of the 2010 IEEE International Conference on Robotics and Automation,Anchorage, AK, USA, 3–7 May 2010; pp. 2493–2498.
19. Zhao, R.; Sidobre, D. Trajectory smoothing using jerk bounded shortcuts for service manipulator robots. In Proceedings of the 2015 IEEE/RSJ International Conference on Intelligent Robots and Systems (IROS), Hamburg, Germany, 28 September–2 October 2015; pp. 4929–4934.
20. Pan, J.; Zhang, L.; Manocha, D. Collision-free and smooth trajectory computation in cluttered environments. *Int. J. Robot. Res.* **2012**, *31*, 1155–1175. [CrossRef]
21. Xu, L.; Cao, M.; Song, B. A new approach to smooth path planning of mobile robot based on quartic Bezier transition curve and improved PSO algorithm. *Neurocomputing* **2022**, *473*, 98–106. [CrossRef]
22. Hu, Y.; Wang, Y.; Hu, K.; Li, W. Adaptive obstacle avoidance in path planning of collaborative robots for dynamic manufacturing. *J. Intell. Manuf.* **2023**, *34*, 789–807. [CrossRef]
23. Fréchet, M.M. Sur quelques points du calcul fonctionnel. *Rend. Circ. Mat. Palermo* **1906**, *22*, 1–72. [CrossRef]

Disclaimer/Publisher's Note: The statements, opinions and data contained in all publications are solely those of the individual author(s) and contributor(s) and not of MDPI and/or the editor(s). MDPI and/or the editor(s) disclaim responsibility for any injury to people or property resulting from any ideas, methods, instructions or products referred to in the content.

Article

A Vision–Language Model-Based Traffic Sign Detection Method for High-Resolution Drone Images: A Case Study in Guyuan, China

Jianqun Yao [1], Jinming Li [1], Yuxuan Li [1], Mingzhu Zhang [2], Chen Zuo [2,*], Shi Dong [2] and Zhe Dai [2]

- [1] CCCC Infrastructure Maintenance Group Co., Ltd., Beijing 100011, China; yaojianqun@126.com (J.Y.); lijm6216@163.com (J.L.); lyx624074264@163.com (Y.L.)
- [2] School of Transportation Engineering, Chang'an University, Xi'an 710064, China; mingzhuzhangzmz@163.com (M.Z.); dongshi@chd.edu.cn (S.D.); zhedai@chd.edu.cn (Z.D.)
- * Correspondence: chenzuo@chd.edu.cn

Abstract: As a fundamental element of the transportation system, traffic signs are widely used to guide traffic behaviors. In recent years, drones have emerged as an important tool for monitoring the conditions of traffic signs. However, the existing image processing technique is heavily reliant on image annotations. It is time consuming to build a high-quality dataset with diverse training images and human annotations. In this paper, we introduce the utilization of Vision–language Models (VLMs) in the traffic sign detection task. Without the need for discrete image labels, the rapid deployment is fulfilled by the multi-modal learning and large-scale pretrained networks. First, we compile a keyword dictionary to explain traffic signs. The Chinese national standard is used to suggest the shape and color information. Our program conducts Bootstrapping Language-image Pretraining v2 (BLIPv2) to translate representative images into text descriptions. Second, a Contrastive Language-image Pretraining (CLIP) framework is applied to characterize not only drone images but also text descriptions. Our method utilizes the pretrained encoder network to create visual features and word embeddings. Third, the category of each traffic sign is predicted according to the similarity between drone images and keywords. Cosine distance and softmax function are performed to calculate the class probability distribution. To evaluate the performance, we apply the proposed method in a practical application. The drone images captured from Guyuan, China, are employed to record the conditions of traffic signs. Further experiments include two widely used public datasets. The calculation results indicate that our vision–language model-based method has an acceptable prediction accuracy and low training cost.

Keywords: vision–language model; traffic sign detection; drone images; multi-modal learning

Citation: Yao, J.; Li, J.; Li, Y.; Zhang, M.; Zuo, C.; Dong, S.; Dai, Z. A Vision–Language Model-Based Traffic Sign Detection Method for High- Resolution Drone Images: A Case Study in Guyuan, China. *Sensors* **2024**, *24*, 5800. https://doi.org/10.3390/s24175800

Academic Editor: George Yannis

Received: 11 July 2024
Revised: 28 August 2024
Accepted: 29 August 2024
Published: 6 September 2024

Copyright: © 2024 by the authors. Licensee MDPI, Basel, Switzerland. This article is an open access article distributed under the terms and conditions of the Creative Commons Attribution (CC BY) license (https://creativecommons.org/licenses/by/4.0/).

1. Introduction

As part of fundamental road infrastructures, traffic signs play a key role in ensuring transportation safety and smooth traffic flow [1,2]. Traffic signs provide essential guidance to restrict the behaviors of vehicles, drivers, cyclists and pedestrians. Numerous vehicle crashes are prevented based on crucial information, including the speed limit and warning of potential hazards. Therefore, the maintaining the conditions of traffic signs becomes an important task in infrastructure maintenance [3]. Regular inspections, cleaning and replacements have a positive effect on the visibility and legibility of traffic signs. In recent years, Unmanned Aerial Vehicles (UAVs) and drones have gained considerable attention in the context of infrastructure maintenance [4–6]. The main advantages of UAVs include their high-resolution images, flying flexibility and high speeds. UAVs can conveniently span extensive areas. Moreover, the combination of UAV images and computer vision techniques has become a prevalent and active topic. In particular, a variety of the traffic sign detection methods are used to analyze UAV images. Inspired by high-performance platforms, the traffic sign detection program provides a low-cost way to identify the locations and categories of traffic signs. Compared with manual operations, inspection time and use of resources are significantly saved.

Traditional traffic sign detection techniques mainly rely on digital-image processing techniques [3]. The following are the three main approaches: color thresholding, shape detection, and feature extraction. Color thresholding techniques transform images into color spaces. Prior knowledge, including hue, saturation, and brightness, is applied to find traffic signs [7]. Shape detection methods employ edge detection, curve fitting and Hough transform to identify the region of interest. The traffic sign area is distinguished from the background. Feature extraction methods concentrate on the local visual characteristics of the image. The program utilizes image descriptors like the Scale-Invariant Feature Transform (SIFT), Histogram of Oriented Gradients (HoG) and Local Binary Pattern (LBP) to generate an image feature vector [8,9]. Next, these feature vectors are fed into a machine learning-based classifier. The relationship between the input feature vectors and the output category is explored by the supervised learning program. The conventional machine learning program includes decision tree, support vector machines and random forest [10,11].

Since 2013, with the rise of deep learning techniques, neural networks and convolutional neural networks (CNNs) are broadly used in a variety of traffic sign detection scenarios [12,13]. In particular, object detection methods have become a prevailing choice for traffic sign detection tasks [14]. As a core component in computer vision, the object detection program focuses on not only pinpointing the location but also classifying the object under research. Thus, there are two main tasks within the object detection framework, namely localization and classification.

In general, the deep-learning-based object detection method can be divided into one-stage and two-stage approaches. As a classical one-stage detection method, You Only Look Once (YOLO) uses a convolutional neural network as the backbone module [15–17]. In order to accelerate the running speed, YOLO directly maps the input image into the output bounding box. Accordingly, the object detection is fulfilled with a single pass of the deep neural network. On the other hand, the two-stage object detection method pays attention to improving the computational accuracy. For example, there are two primary steps in Region-based Convolutional Neural Networks (RCNNs) [18–20]. The first step is to generate numerous candidate bounding boxes. Then, the computer concentrates on classifying these candidate regions. Compared with the one-stage method, the two-stage approach alleviates the misdetection issues and achieves high detection accuracy. Aiming at achieving real-time detection, a collection of lightweight networks can reportedly speed up traffic sign detection. Attention mechanisms and feature aggregation modules are advisable methods for improving efficiency [21–23]. In recent years, Detection Transformer (DETR) networks have emerged as a cutting-edge method in the field of object detection [24]. The main contribution of DETR networks is to combine the local visual feature extraction capabilities of convolutional neural networks and the sequence data analysis strength of the Transformer model. In other words, DETR networks not only consider the association between each bounding box and the target instance but also account for the positional relationships between individual bounding boxes. Therefore, the detection quality issue is effectively mitigated.

With the advancements in computation platforms and image acquisition, deep-learning-based traffic sign detection methods have gained a wide range of real-world applications. However, the extensive utilizations of the deep learning techniques encounter several substantial challenges. First, deep neural networks such as YOLO, RCNNs, and DETR networks highly rely on labeled image datasets to learn the non-linear projection between input images and output predictions. In real-world scenarios, it is costly and expensive to manually annotate high-value targets in numerous images. The lack of high-quality datasets not only slows down the convergence of neural networks but also results in the overfitting problem. Second, training and inference are two main steps within the deep neural networks. The performance of the object detection program is considerably impacted by the difference between training materials and real-world testing images. In practice, traffic signs encompass a variety of categories and rich semantic information. It is challenging to build a training dataset covering all potential cases and train a neural network to handle unseen categories. Third, both YOLO and RCNNs use convolutional neural networks as a fundamental component for extracting visual features. Hyperparameter settings, such as learning rate, gradient descent and the number of convolutional layers, have a considerable influence on the performance of neural networks. The parameter specification step often consumes a lot of time and computational resources in the real-world scenario.

Since 2020, vision language models (VLMs) have gained significant attention in the community of artificial intelligence and computer vision [25]. The benefits of VLMs are listed as follows: (1) As a core concept in VLMs and large-scale pretrained models, the self-supervised learning technique is dedicated to directly extracting visual features from images without manually labeled data [26,27]. In training with massive image datasets, the self-supervised techniques significantly reduce the resource consumption associated with image collection and manual labeling. The neural network pretrained by the self-supervised learning technique becomes a favorable starting point for tackling the downstream image analysis tasks. (2) VLMs have a strong multi-modal capability and zero-shot prediction ability. The traditional image classification and object detection programs are highly dependent on discrete labels to express target instances. In the VLM framework, the similarity between visual content and textual explanation can be quantified by comparing feature vectors. The multi-modal ability allows users to customize text descriptions to explicitly explain the semantic meaning of target instances. (3) Vision transformers (ViTs) are broadly used as the basic architecture in most VLMs [28–30]. Compared to CNNs, a ViT offers a better image understanding ability and analysis accuracy. The main reason lies in its capability to exploit long-range spatial dependency and global features in input images. Thus, a ViT excels in complex scenarios such as image classification, semantic segmentation and image generation.

In this paper, we explore a vision–language model-based traffic sign detection method. Our research focuses on addressing the training cost issue and slow deployment problem. Within the existing traffic sign detection framework, considerable computation resources are consumed by creating human annotations and fine-tuning a neural network. In addition, the inconsistency between training sets and testing sets can cause difficulties in program deployment. Compared with previous programs, two key points of the proposed method are as follows: (1) As important prior knowledge, the text description is used to depict the traffic signs and guide the image processing program. (2) To solve the training burden, we apply the large-scale pretrained network to analyze the image and text input. The basic steps of our method are listed as follows. First, a keyword dictionary is created to express the semantic meaning of various traffic signs. We referred to the Chinese national standard to obtain the shape and color information of common traffic signs. In addition, the Bootstrapping Language-image Pretraining version 2 (BLIPv2) model is applied to realize the image translation. High-quality text descriptions are automatically produced in accordance with the representative traffic sign images. Second, Contrastive Language-image Pretraining (CLIP) is launched for multi-modal feature extraction. An image encoder is responsible for extracting visual features from UAV images. In comparison, a text encoder is devised to project each element of the keyword dictionary into a high-dimensional language feature vector. Third, we calculate the similarity according to the cosine distance between the visual and textual features. The Softmax function is applied to predict the probability distribution and determine the category of each traffic sign instance.

We test the proposed method in a highway scenario in Guyuan, China. Accuracy, precision, recall, F1 score and frame per second (FPS) are used to quantitatively evaluate the performance. Furthermore, the performance of the VLM is checked by two public traffic sign datasets. The computation results indicate that our method achieves high classification quality with low resource consumption. Considering its ability, the VLM offers new insights for traffic sign detection research.

The main contributions of our proposed method are summarized as follows:

(1). The multi-modal learning technique is introduced to alleviate the reliance on the annotations and solve the domain shift issue. Instead of discrete labels in the annotated dataset, the text description becomes an important material in regard to representing the prior knowledge of traffic signs.

(2). In order to save training time, we launch the vision–language model to analyze UAV images as well as text descriptions. There is no fine-tuning step because our program benefits from the strong generalization ability of the large-scale pretrained model. Therefore, the proposed method can be efficiently deployed in a new scenario.

(3). The proposed method is applied in a real-world case study in Guyuan, China. Two public datasets are used to compare our program with other milestone methods. The computational results indicate that the proposed vision–language model-based method exhibits competitive performance in terms of generalization ability and cross-domain adaptation. Exploring the multi-modal learning and large-scale network methods may prove to be beneficial in regard to solving the practical traffic sign detection task.

The rest of the paper is arranged as follows: Section 2 provides a brief overview of the previous deep learning methods in the field of traffic sign detection. The technical details of the proposed program are explained in Section 3. Sections 4 and 5 introduce the practical applications of our method. A group of evaluation metrics are presented to assess the detection performance. Finally, the conclusion is summarized in Section 6.

2. Existing Deep-Learning-Based Traffic Sign Detection Method

As a mainstream method in the field of the traffic sign detection, the YOLO neural network is renowned for its excellent inference speed and real-time performance [15]. Figure 1 provides the basic steps of the YOLO-based traffic sign detection program. At first, the researchers and participants depict the bounding boxes on each traffic sign instance. A training image dataset associated with human annotations is created. Then, a training program is launched to train the YOLO network. The YOLO network concentrates on learning the relationship between the input image and output boxes. Finally, the performance of the YOLO network is examined in a practical application. The traffic sign instances are rapidly found in the real-world case. To save computational time and resources, YOLO employs a convolutional neural network to simultaneously realize the localization and recognition tasks. On one hand, the object localization is equivalent to a regression problem. The core goal is to predict not only the coordinates of the object center but also the height and width of the bounding box. On the other hand, the program treats the recognition task as a classification program. The neural network focuses on estimating the possibility of each category.

To further improve the capability of YOLO, researchers continuously report new technical improvements. In 2017, the YOLOv2 network is suggested. Data augmentation, the smoothing L1 loss function and Softmax function are introduced to improve the detection accuracy [31]. To address the concern of the bounding box size, YOLOv2 uses a clustering algorithm to produce anchor boxes from the training dataset. Subsequently, the YOLOv3 network adopts the intersection of union (IoU) loss function and the Focal loss function to enhance object localization and classification quality [32]. To accurately analyze images, YOLOv3 applies Darknet-53 as its backbone module. The increased number of convolutional layers has a positive influence on capturing complex visual information. In 2020, YOLOv4 and YOLOv5 focus on extracting hierarchical image features [33,34]. The path aggregation network, spatial pyramid pooling and feature pyramid network are reported to integrate feature information across different resolutions. These modifications enable YOLO to simultaneously detect large and small objects. In the past three years, YOLOv6, YOLOv7, and YOLOv8 have been proposed [35,36]. Cross-scale training, attention mechanisms and model distillation are incorporated into the YOLO framework. These developments significantly advance the capabilities of one-stage target detection networks and greatly improve the ability of neural networks to detect target instances.

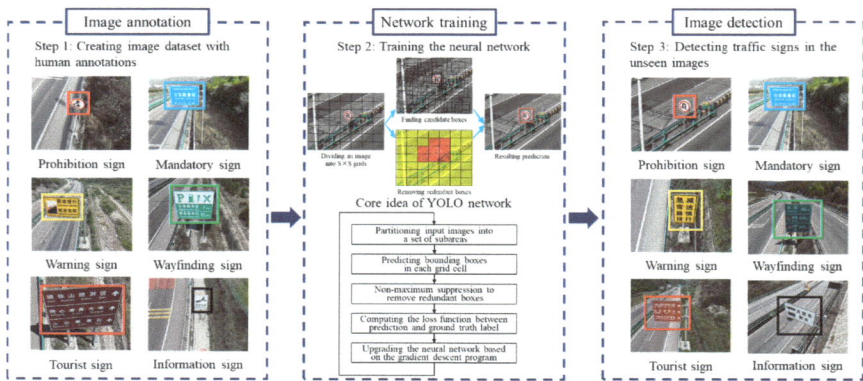

Figure 1. Basic workflow of YOLO-based traffic sign recognition.

Although the YOLO network has made significant progress in the field of intelligent transportation, the traffic sign detection methods encounter the following several key technical limitations: (1) The mainstream traffic sign detection methods are highly dependent on labeled image datasets. The number of images, the diversity in the dataset and the annotation quality are decisive factors in the recognition accuracy. In practical applications, it is challenging to collect an adequate amount of labeled data for specific scenarios. The lack of training data slows down the neural network convergence and contributes to the overfitting problem. (2) There are two main procedures in object detection: training and inference. Therefore, one key challenge is to tackle the difference between the training dataset and the actual measured dataset. Traffic sign instances have a broad range of styles, colors and shapes. It is difficult to introduce abundant semantic information and appropriate diversity into the training image dataset. (3) Networks such as YOLO and RCNNs use convolutional neural networks as their core backbone module. Hyperparameter settings, including learning rate, optimizer, neural network architecture and image resolution, have a substantial effect on the recognition accuracy. Researchers have to spend considerable time to find the optimal parameter configuration for specific application scenarios. The extensive time requirement brings a difficulty to the rapid and flexible deployment of traffic sign detection programs.

3. The Vison Language Model-Based Traffic Sign Detection Method

To realize the rapid deployment of the traffic sign detection, this paper introduces a vision–language model-based approach to deal with high-resolution UAV images. As Figure 2 shows, the proposed method consists of three basic modules. At first, a keyword dictionary is constructed utilizing the Chinese national standard and BLIPv2. Next, we carry out the multi-modal feature extraction. CLIP is activated to generate the visual and textual feature vectors. Finally, the program performs the similarity measure step to determine the category of traffic signs.

There are two benefits of our method, as follows: (1) The dependence on the annotated dataset is significantly mitigated by the multi-modal learning program. The text description becomes a valuable alternative to the discrete image labels. (2) The fine-tuning step is saved by the employment of vision–language models. The pretrained VLM programs create a solid foundation to analyze the UAV images as well as text descriptions.

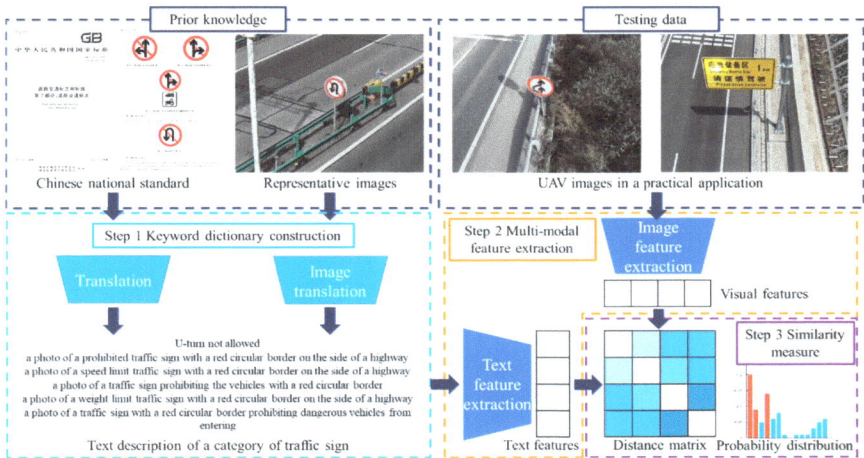

Figure 2. Basic workflow of the vision–language model-based traffic sign detection approach. The Chinese national standard of traffic sign can be found in Reference [37].

3.1. Keyword Dictionary Construction to Explain Traffic Signs

The first step of the proposed method in this work is to build a keyword dictionary describing the semantic information of various traffic signs. To generate high-quality text descriptions, we design two approaches in the dictionary construction step. As the first method, it is important to note that the visual characteristics of all traffic signs are strictly constrained with the national regulation. According to Chinese national standards [37], there are six main categories of traffic signs. Each class has its specified combination of the base color, character color, border color, shape and location. Representative images for each category are shown in Figure 3.

Figure 3. Six categories of Chinese traffic signs.

The specific definitions of each type of traffic sign are discussed as follows. (1) Prohibition signs are used to impose strict restrictions on traffic behaviors. Common examples include stop, no vehicles, no pedestrians and no U-turn. Prohibition signs are usually circular in China. In order to provide favorable visibility, the main colors include red, white and black. (2) Mandatory signs are employed to guide drivers as well as pedestrians. Typical examples include turning signs, roundabouts, one-way streets and pedestrian crossings. Mandatory signs are characterized by their circular or rectangular shape, blue background and white symbols. (3) Warning signs promote traffic participants to observe road and traffic conditions. Sharp curves, continuous curves, rockfall areas and crosswinds are common examples of warning signs. In general, triangular shapes with large yellow areas and black characters are used to indicate the presence of dangerous situations. (4) Wayfinding signs provide information on road direction, location and distance. They usually have large areas to convey adequate information. Rectangular borders and white text are main features. The blue and green backgrounds are separately used to imply urban roads and highways. (5) Tourist area signs are responsible for supplying directions and distances to tourist attractions. To improve visibility, they have brown backgrounds and white fonts. (6) Information signs focus on providing information about off-road facilities, safe driving and other relevant information. Typical examples include motorway numbers and seat belt reminders. The main shape is rectangular with white areas and black characters.

Based on the text description in Chinese national standards, we construct an ensemble of English sentences to explain various traffic signs. Figure 4 illustrates English descriptions corresponding to each traffic sign category. In general, five or six sentences are summarized from the explanation in Chinese national standards. It is clear that the color and shape information plays an important role. For example, the phrase 'a red circular border' is commonly used to describe the prohibition sign. In general, red is a signal to imply a dangerous situation and offer high visibility. Circles are beneficial in terms of improving clarity and simplicity. In comparison, the words 'blue', 'yellow', 'green', 'dark red' and 'white and black' are broadly applied to describe the rest of traffic signs. Furthermore, a noteworthy phenomenon is that only three sentences are created to describe the tourist sign. The main reason for this lies in the limited content of the Chinese national standard. Compared with the other five categories, the regulation of tourist signs is not adequate.

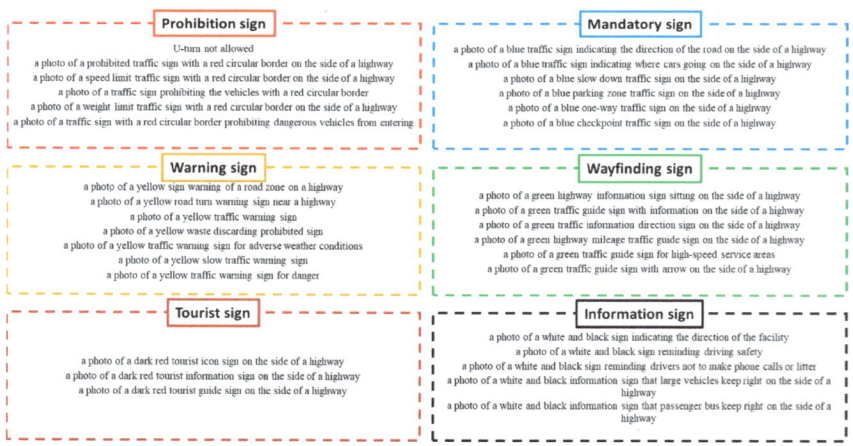

Figure 4. Keyword dictionary constructed by Chinese national standards.

Moreover, we introduce a second way to create the keyword dictionary. Compared with the text description in the Chinese national standard, the example UAV images become explicitly prior material to express the visual characteristics of traffic signs. As Figure 5 shows, we activate the Bootstrapping Language-image Pretraining Version 2 (BLIPv2) program to convert an image into an English description [38,39]. As an excellent vision–language pretraining method, BLIPv2 is devised to connect the objects in images and their language descriptions. Aiming at improving the performance, BLIP applies a unified multi-task learning framework to simultaneously fulfill the understanding-based task as well as the generation-based task.

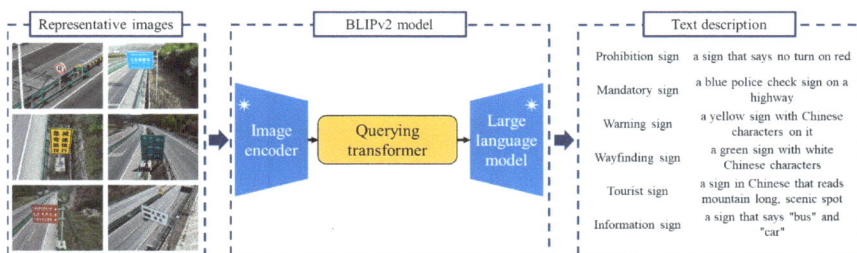

Figure 5. Keyword dictionary constructed by representative images and BLIPv2 model. The mark * indicates that the frozen encoder networks are used by BLIPv2.

There are three primary objectives within the BLIP training program. First, the image–text contrastive learning module focuses on aligning the image representation and text representation in the feature space. As a contrastive learning step, the program is devoted to maximizing the image–text similarities between semantically matching samples across modalities. In comparison, the distance between the image and text is supposed to be large when unrelated instances are provided. Second, the image-grounded text generation step is responsible for generating the text description according to the intrinsic characteristics of an input image. The visual information is converted into the image caption. Using the cross-entropy as the loss function, the text generator attempts to enlarge the likelihood of each word in an autoregressive manner. The label smoothing strategy is utilized to improve the robustness and generalization of the neural network. Third, the image–text matching module is suggested to learn the fine-grained alignment between image and text representation. As a binary classification problem, the image–text matching program is suggested to predict whether an image–text pair is matched or unmatched. The similar instances constitute a positive pair. In comparison, a negative image–text pair indicates that there are two unmatched and incompatible samples. With the intention of developing the prediction accuracy, a bi-directional self-attention mask is employed to bridge images and texts. A two-class linear classifier is used to output the classification result. In addition, several technical modifications are made by BLIPv2 to develop the training efficiency and reduce trainable parameters. A lightweight Querying transformer is devised to bridge the modality gap and facilitate cross-modal interaction. Moreover, BLIPv2 highlights the role of large-scale pretrained neural networks. On one hand, a frozen image encoder network is applied to yield the high-level visual feature in accordance with an input image. On the other hand, the generative language capability is improved by a frozen large language model.

Based on the pretraining step in a massive dataset of images and their captions, BLIPv2 is capable of understanding the visual information as well as generating high-quality text descriptions. In this work, we employ a BLIPv2 program to analyze the traffic signs captured by high-resolution UAV images. In order to avoid the influence of outliers, we assign two representative images into each category. Since there are six classes of Chinese traffic signs, 12 representative images are involved in this case. The English sentences created by BLIPv2 are shown in Figure 5. It is apparent that the color information plays an important role in text description. The words red, yellow, blue and green explicitly present the main characteristics of Chinese traffic signs.

3.2. Multi-Modal Feature Extraction with Contrastive Language-Image Pretraining

After the keyword dictionary construction step, the next technical problem is to analyze the drone image and text description. It is necessary to convert the unstructured image and language data into numerical vectors. The feature vector is also referred to as embedding in the deep learning community. In the feature space, each vector represents an instance. The semantically matching samples share similar positions. In contrast, the large distance between the two points reveals that there is a substantial difference between the two examples.

Image and text feature extraction programs are active topics in the field of computer vision and natural language processing. The traditional image processing technique introduces a variety of handcrafted descriptors to analyze the input image. For example, Scale-invariant Feature Transform (SIFT) and Histogram of Gradient (HoG) pay attention to calculating the distribution of local gradients. The difference between the template center and neighboring points becomes a decisive element with the local binary pattern (LBP) framework. With the development of deep learning, the neural network becomes a valuable alternative to achieve image feature extraction. After training on a massive dataset, convolutional neural networks (CNNs) and vision transformers (ViTs) have a powerful capability for capturing task-specific and robust embeddings. A high-dimensional feature vector is produced to explain the visual content. On the other hand, the text characterization technique also has experienced the shift from handcrafted descriptor to deep-learning-based method. The traditional text feature extraction techniques include bag-of-words (BoW), term frequency–inverse document frequency (TF-IDF) and N-grams. The frequencies of specific words have a huge influence on feature vectors. Inspired by the success in sequential data processing, recurrent neural networks (RNNs), long short-term memory (LSTM) and transformer networks are broadly used to extract high-level features from the sentences.

In this work, we apply the Contrastive Language-image Pretraining (CLIP) framework to realize the image and text feature extraction [40]. As a prominent method in the field of multi-modal learning, the core concept of CLIP is to create a joint representation space for both images and text. In other words, both visual information and textural descriptions are projected into a unified feature space. The advantages of CLIP are embodied by the following three aspects. (1) A transformer network is used to capture the high-level features from the raw data. There are two major components within CLIP. On one hand, an image encoder pays attention to fulfilling the image feature extraction. A ViT model is applied to project a 2D input image into a 1D feature vector. Rather than the convolution calculation in a CNN, a ViT applies the attention mechanism to explain the interaction between different image patches. Therefore, a ViT is a viable tool for exploring the global structure and extract visual content. On the other hand, the text encoder is responsible for mapping textual descriptions into numerical representations. A transformer-based architecture is utilized to capture long-range dependency within language sentences. The self-attention mechanism in the transformer plays an important role in understanding the correlation between each word. (2) Self-supervised learning is an essential part of the CLIP framework. Aiming at alleviating the reliance on data labeling, there is no human annotation in the training dataset. To promote the network training program, CLIP is dedicated to automatically generating labels of training examples. Motivated by the success of contrastive learning, the relationship between images and text becomes a key concept. On the basis of a massive dataset of image–text pairs, CLIP trains neural networks by encouraging the similar instances to be close together. The image and its corresponding caption share the close positions in the feature space. In comparison, incompatible image and text examples are pushed apart and assigned a large distance. (3) There are 400 million image–text pairs in the training dataset. The diverse examples in this extensive dataset establish a favorable foundation to handle the overfitting problem. Instead of memorizing specific instances, the neural network tends to capture the underlying patterns within the training data. Moreover, a wide range of training data is useful for ensuring generalization and robustness. The image–text pairs with varying conditions help the program to mitigate the data bias and generate an effective representation of different modality.

Based on the preceding discussion, one important procedure is generating visual feature vectors in accordance with the CLIP image encoder network. An illustration is shown in Figure 6. A pretrained ViT program is launched to analyze the drone image. There are four key steps, as follows: First, the input image is evenly partitioned into several non-overlapping patches. In other words, a sequence of image patches is provided to the following steps. In general, a patch of size 16 × 16 or 32 × 32 is used. Second, the program converts each 2D patch into a low-dimension vector on the basis of a reshaping operation and a linear projection layer. This vector is also referred to as patch embedding in many studies. Moreover, the position embeddings are combined with the patch embedding in order to record the location of each patch. Third, a transformer encoder module is designed to deal with a sequence of patch embeddings. While encoder-decoder architecture is widely used in the transformer for natural language processing, a ViT only applies the encoder module to generate the image embedding according to the visual content. Two basic modules are involved in the transformer encoder step. On one hand, the self-attention mechanism concentrates on exploring the correlation between each element in the embedding sequence. On the other hand, a feed forward network conducts non-linear calculations to learn complex structures. Fourth, the multi-head attention module becomes a sensible choice in a ViT in regard to learning a variety of spatial dependencies between image patches. Based on a sequence of patch embeddings, several self-attention modules are carried out in parallel. The outputs of the attention modules are concatenated and fed into the final layer to generate a rich visual representation.

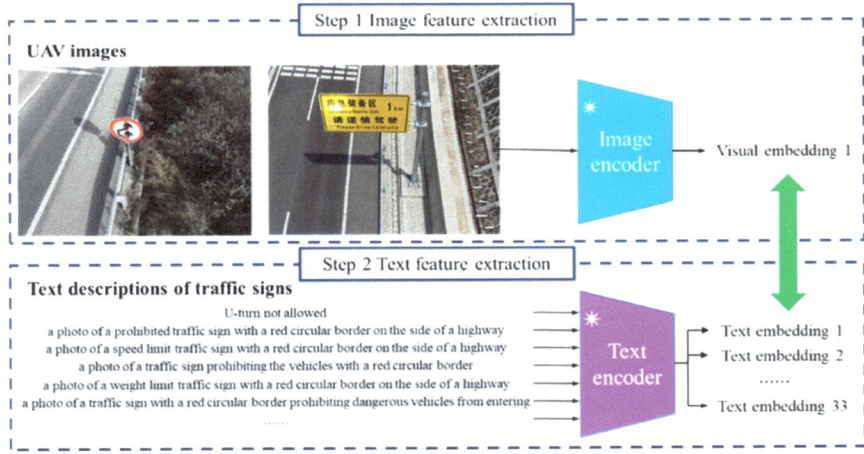

Figure 6. High-dimensional feature extraction with CLIP network. The mark * indicates that we adopt the frozen encoder networks to realize feature extraction.

Moreover, our program launches a transformer-based text encoder network to realize the text feature extraction. Similar to a ViT, the attention mechanism plays an important role in the transformer network [41]. The second row of Figure 6 provides the basic steps of the text encoder network. The technical details of the text encoder network are discussed in the following. At first, tokenization is a necessary preprocessing step. The computer divides the input sentence into individual words. The starting and ending markers are predefined to express the structure of a sentence. Next, a pretrained word embedding program is carried out to map the English words into a numerical vector. This vector conveys the basic semantic content of each word. Afterward, the program uses the positional embedding method to preserve the location of each word. Then, a transformer encoder network is performed to quantify the correlation between each two words. Like a ViT, multi-head attention and feed forward networks are essential components in the transformer architecture. Finally, a linear projection layer is launched to generate the resulting output. The embeddings independently created by the attention modules are combined to produce an effective text feature vector.

3.3. Traffic Sign Classification with Similarity Measurement

By understanding both the visual information of the drone image and the semantic information of the text description, the computer is able to determine the category of traffic signs. The predicted probability distribution is generated by calculating the similarity between the visual feature vector and the text feature vector. In this paper, our program uses cosine distance as the criterion for similarity calculation. Different from Euclidean distance and Hamming distance, cosine distance focuses on the angle between the input vectors. The influence of vector magnitude is significantly alleviated on the similarity calculation. Therefore, the cosine distance becomes a suitable approach to quantify the similarities between high-dimensional vectors. In addition, the value of cosine distance ranges from 0 to 1. The value of 0 indicates that two input vectors share the same direction. Assuming that we have n-dimensional visual feature vector X and textual feature vector Y, the cosine distance between the two embeddings is defined as follows:

$$\text{dis}(X, Y) = 1 - \frac{\sum_{i=1}^{n} x_i y_i}{\sqrt{\sum_{i=1}^{n} x_i^2} \sqrt{\sum_{i=1}^{n} y_i^2}} \tag{1}$$

where x_i is the i-th element of the feature vector X.

Based on the calculation results of the cosine similarity, our method uses the Softmax function to generate the predicted probability distribution. Based on the discussion in Section 3.1, we produce 33 keywords according to the Chinese national standard. Suppose that the computer focuses on addressing one input image. Thus, one visual feature vector and 33 textual feature vectors are created in this case. In accordance with Equation (1), a distance set $(dis_1, dis_2, \ldots, dis_{33})$ is yielded. Then, the predicted probability corresponding to the k-th text description is calculated as follows:

$$p_k = \frac{e^{dis_k}}{\sum_{j}^{33} e^{dis_j}} \tag{2}$$

On the basis of the probability distribution, our program selects the text description associated with the maximum probability as the predicted category.

4. The Traffic Sign Detection Task: A Case Study in Guyuan, China

4.1. Data Collection and Data Preprocessing

In this section, a practical application in Guyuan, China, is used to examine the proposed method. The location of Guyuan is shown in Figure 7. As a city in Northwestern China and Ningxia Province, Guyuan spans over 10,540 square kilometers and has a residential population of 1,142,000 people. Moreover, Guyuan is characterized by a complicated topography, drought climate and low rainfall. Affected by the Jing River and Qingshui River, the landscape of Guyuan consists of large interlaced loess, beams, mountains and trenches. The highway network of Guyuan extends for about 408.6 km. In particular, a highway of 333.6 km in length has been constructed over the past three years. The major nation-level highways include G70 Fuyin, G85Yinkun and G22 Qinglan. Due to the challenging natural environment and a high proportion of bridges and tunnels, infrastructure maintenance in Guyuan has become a difficult task. Therefore, it is crucial to develop an effective detection method that continuously monitors the condition of traffic signs. With the intention of ensuring transportation safety as well as operation efficiency, the damaged signs should be promptly identified and repaired.

In this application, we employ a UAV to investigate the infrastructure condition of highways in Guyuan. An amount of 111 high-resolution images of size 8000 × 6000 pixels are captured to indicate the operating state of the traffic signs. Figure 3 provides representative examples of the traffic signs. It is apparent that there are a variety of imaging circumstances in our UAV dataset. The imaging angles, shooting distance and illumination condition considerably enlarge the diversity of traffic sign datasets. These variations present significant challenges for the subsequent classification programs.

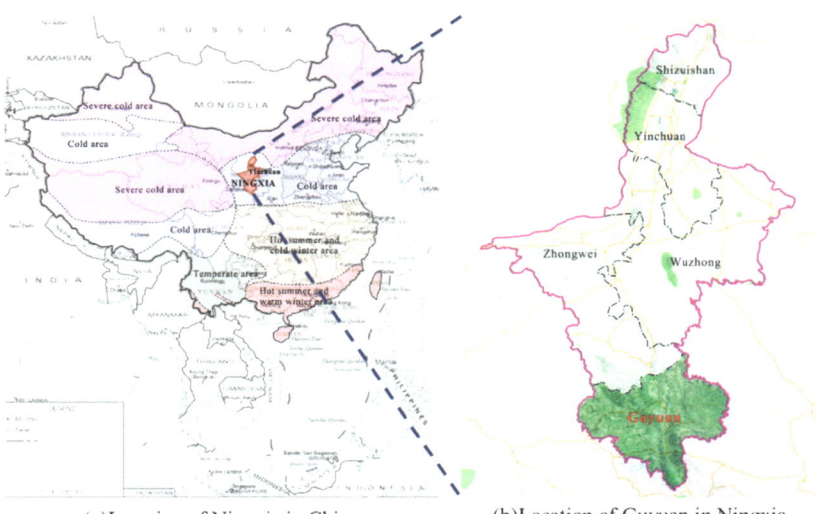

(a) Location of Ningxia in China (b) Location of Guyuan in Ningxia

Figure 7. Locations of Guyuan in China.

4.2. Traffic Sign Detection with VLM and Supervised YOLO Networks

For the 111 UAV images, this study utilizes the CLIP image encoder to extract visual features. Given an input image, the image encoder creates a high-dimensional feature vector. The ViT-B/32 model is chosen as the backbone module. As an important component within the CLIP framework, there are three main parts of ViT-B/32, namely the input embedding layer, the Transformer encoder and the multi-layer perceptron. In ViT-B/32, the total number of learnable parameters is approximately 86 million. The image processing program is explained in detail. (1) The input embedding layer is responsible for partitioning the input image into a series of patches of size 32×32. Based on linear mapping operation, the image patches are encoded into 768-dimensional vectors. (2) The transformer encoder concentrates on finding long-range dependencies between high-dimensional vectors and exploring the spatial structures of the image. Furthermore, a multi-head attention mechanism is applied to identify the spatial relationship from various perspectives. Aiming at capturing the fine-grained visual content, 12 encoder modules are adopted by ViT-B/32. (3) As the final component, the multi-layer perceptron is devoted to controlling the dimensionality of the output vector. The program creates a 512-dimensional feature vector to represent the input UAV image. The detailed technical specifications of the ViT-B/32 model are available in the literature [29]. Since 111 UAV images are involved, the CLIP image encoder associated with ViT-B/32 model yields a feature matrix of size 111×512 to express the visual content.

Next, the CLIP text encoder is launched to analyze text descriptions created by the Chinese national standard as well as the BLIPv2 model. In this application, the Text Transformer is selected as the backbone network. There are 12 layers within the Text Transformer framework. Each layer has eight multi-head attention mechanisms. Nearly 63 million parameters are involved in this deep neural network. The architecture of the Text Transformer network is detailed in the literature [40,41]. It is worth noting that there are two ways to generate text descriptions in our method. On one hand, 33 sentences are created according to the Chinese national standard. Therefore, a text feature matrix of size 33×512 is obtained by the CLIP text encoder. On the other hand, our program views 12 representative images as the prior material to express the six categories of Chinese traffic signs. BLIPv2 is activated to translate images and produce the text description. Therefore, our program yields a text feature matrix of size 12 × 512.

Based on the feature matrix mentioned above, we calculate the cosine distance as a measure of the cross-modal similarity. The class probability estimation is performed using the Softmax function. In the following paragraph, the drone images in Figure 3 and the text description in Figure 4 are used as the example. The cosine distance between the visual feature matrix and the text feature matrix is shown in Figure 8. The three smallest distances are highlighted in orange. Figure 9 displays the predicted probability distributions. The maximum probability and its corresponding category are emphasized in red.

In order to provide an in-depth understanding, the computation result in Figures 8a and 9a is analyzed. The three keywords with the highest probabilities are "U-turn not allowed", "a photo of a prohibited traffic sign with a red circular border on the side of a highway" and "a photo of a traffic sign prohibiting vehicles with a red circular border". It is clear that the first keyword describes the prohibition of U-turn behavior. In comparison, the second and third keywords emphasize the red boundary of the prohibition sign. Figure 9a presents the predicted probability of the traffic sign category for this image. The possibilities associated with three candidate keywords are 20%, 14% and 9%, respectively. Accordingly, the computer identifies the target category in Figure 7a as a prohibition sign. The calculation results in Figures 8 and 9 indicate that the vision–language model method used in this study effectively predict the categories of traffic sign images.

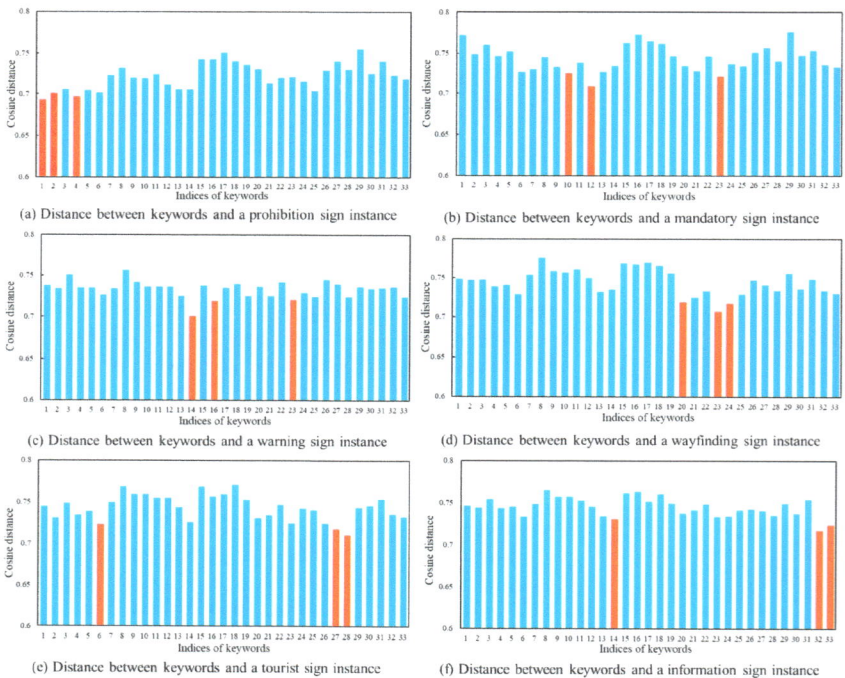

Figure 8. Cosine distance between traffic sign images and keywords.

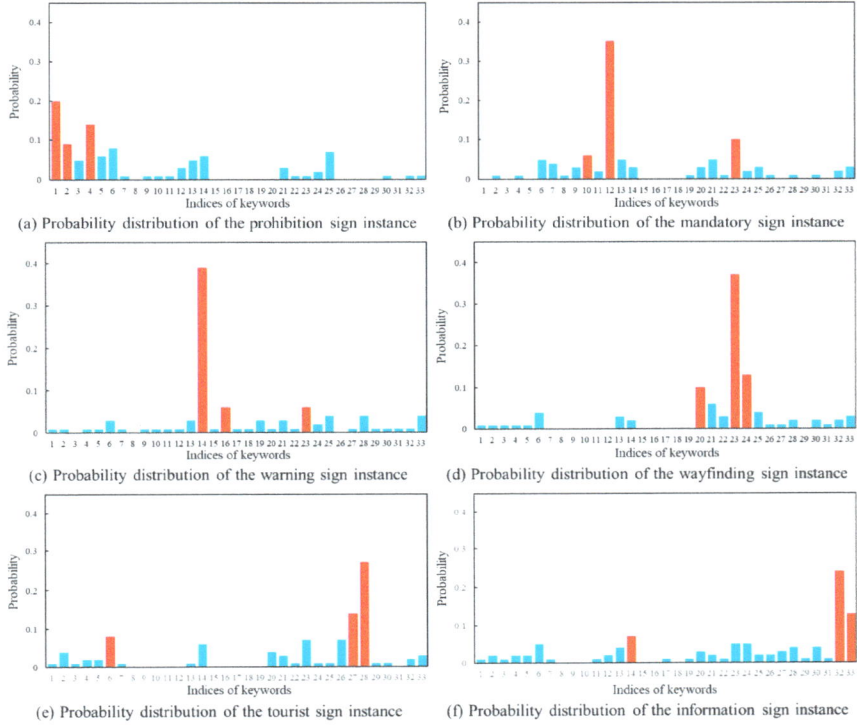

Figure 9. Predicted probability distribution of traffic sign images.

To further evaluate the performance of the designed method, we manually classified all 111 traffic sign images. Based on the human identification, the drone images consist of 23 prohibition signs, 5 mandatory signs, 26 warning signs, 35 wayfinding signs, 13 tourist area signs and 9 information signs. Next, we applied the proposed method to classify the traffic signs in the UAV images. The accuracy and reliability of the recognition method are quantitatively assessed using classification accuracy. In the classification task, the ratio between the number of correct prediction and the total number of samples becomes the evaluation metric. Based on the text description created by the Chinese national standard, Figure 10a displays the number of samples as well as the classification recognition accuracy for each category. The overall classification accuracy is 86%. A noticeable phenomenon is that there are no human annotations in our detection program. During the keyword construction step, the Chinese national standard plays a key role in generating the text description. The drone images are not involved. Therefore, the annotation cost is significantly mitigated by the keyword dictionary.

Next, we analyze the classification results performed by comparing the drone images and text descriptions created by BLIPv2. Given two representative images for each category, BLIPv2 creates 12 English sentences to summarize the main characteristics of Chinese traffic sign. Then, the CLIP text encoder network is carried out to generate the text embeddings. Finally, the classification task is realized by the cosine distance and the Softmax function. The text description with the maximum probability becomes the classification result. The classification result is shown in Figure 10b. Given 111 drone images, the classification accuracy is 89%. This finding indicates that the multi-modal ability of the vision–language model establishes a robust foundation for the traffic sign detection problem.

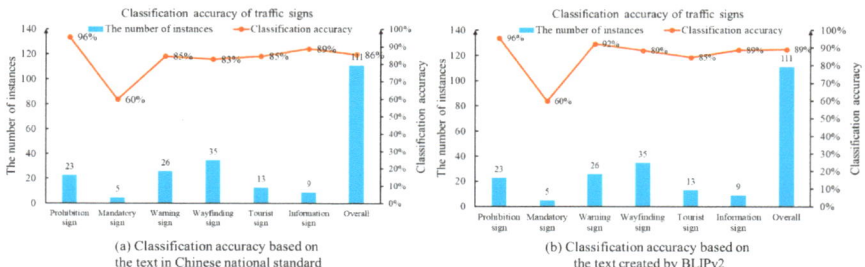

(a) Classification accuracy based on the text in Chinese national standard

(b) Classification accuracy based on the text created by BLIPv2

Figure 10. Classification accuracy of traffic sign images based on the proposed method.

The computation time for each computation module is discussed in Figure 11. We compile our traffic sign detection program in Python 3. As a prevailing programming library, PyTorch 2.3 is used to implement the deep learning method. Our computer configuration includes Windows 10, a Inter I5-12490F of 3.0 GHz, and a memory of 32 GB. A NVIDIA RTX 4060 is employed to carry out the deep learning computation. It should be noting that we have two keyword dictionaries. In accordance with Figure 4, the first dictionary is created using the Chinese national standard. In comparison, 12 text descriptions are generated by the BLIPv2 program.

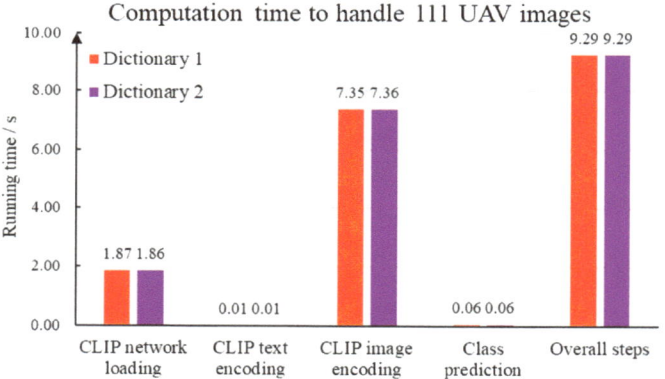

Figure 11. Time consumption of each component in the vision–language model-based method.

The detailed time consumption is explained as follows. First, the BLIPv2 program attempts to translate 12 representative images into text descriptions. Based on our computing platform, 6.13 s are spent to handle one image. Thus, the dictionary construction step takes 73.56 s to generate appropriate keywords. Second, 1.86 s are consumed to load the multi-modal processing program within the CLIP framework. Third, the computer takes 0.014 s to extract high-dimensional embedding from the 33 sentences generated from the Chinese national standard. In comparison, 0.006 s is necessary to analyze 12 BLIPv2 text descriptions. Fourth, the CLIP image encoder network is launched to extract visual features from a given image. Since there are 111 images in our dataset, 7.35 s are required to create a high-dimensional feature matrix. Therefore, CLIP takes 0.06 s to analyze high-resolution drone images. Fifth, the proposed method utilizes cosine distance and the Softmax function to specify the category of each traffic sign. A time of 0.06 s is taken to deal with 111 drone images. Based on the discussion mentioned above, the overall running time of our vision–language model-based traffic sign detection is 9.29 s.

According to Figure 11 and the preceding time analysis, it is obvious that the high computation efficiency is realized by introducing the vision–language model in the traffic sign detection task. Accordingly, the advantages of the multi-modal traffic sign detection method proposed in this paper include the following: (1) The multi-modal recognition approach significantly reduces the running cost for image labeling and image dataset building. The keyword dictionary is rapidly created based on the Chinese national standard and representative images. In comparison, a high-quality labeled dataset is an important part of the mainstream deep learning frameworks. Substantial labor is consumed to create accurate human annotations. Moreover, noise in the image annotations has a negative effect on the network training procedure. (2) Our vision–language model-based method directly employs a pretrained neural network to fulfill the image and text feature extraction. The need for training and fine-tuning deep neural networks is mitigated to realize the rapid deployment. The pretrained network significantly alleviates the dependence of the deep learning method on high-performance computing resources.

4.3. Comparison with Supervised YOLO Networks

To quantitatively assess the performance of the designed method, we implemented YOLOv5 and YOLOv8 networks as the comparison methods. YOLOv5 is an object detection network developed by Ultralytics in 2020 [34]. Compared to the previous methods, YOLOv5 introduces the Focus loss function, path aggregation network and feature pyramid with the intention of developing detection accuracy and efficiency. On the other hand, YOLOv8 is released by Ultralytics in January 2023 [35]. The improved computational accuracy and efficiency is achieved across general-purpose object detection datasets.

As a classical supervised object detection network, the annotated bounding boxes are necessary materials within the YOLO network. In this case, all traffic signs in 111 drone images are identified by the expert. Then, our program divides 111 images into training and testing sets using an 80% to 20% ratio. In other words, the first 88 images are independently adopted to train two YOLO neural networks. The remaining 23 images are used to assess the performance of the deep neural network. To ensure reliable prediction results, we set the number of training epochs as 150 and the batch size for the neural network as 4. The network optimizer and learning rate schedule separately specify the adaptive moment estimation (Adam) and the cosine annealing schedule. The initial learning rate is controlled as 0.001. Based on the dataset configuration mentioned above, YOLOv5 takes 28.2 min to complete the neural network training process. In contrast, the training program in YOLOv8 took 28.8 min. In accordance with the time performance, it is evident that supervised learning frameworks heavily rely on labeled datasets. The neural network is constantly upgraded based on the visual features of a specific dataset. However, the image labeling and neural network training are time-consuming operations.

The trained YOLO models are employed for the traffic sign detection task. Treating 23 images as the testing dataset, the YOLOv5 network correctly predicts 20 traffic signs. However, two traffic signs are not identified. One traffic sign is misclassified. Therefore, the overall classification accuracy of 23 testing images is 86%. On the other hand, 19 traffic signs are accurately classified by the YOLOv8 program. Three traffic signs are neglected and one traffic sign instance is misclassified. Thus, the classification accuracy of YOLOv8 is only 83%. The classification accuracy of the two YOLO networks for various traffic signs is explained in Figure 12. It is evident that two object detection networks suffer from the limited accuracy of mandatory signs, tourist signs and information signs. Only 50% of the classification accuracy is realized when two mandatory signs are fed into the two YOLO networks. The reason for the limited accuracy of the YOLO networks lies in its dependency on the labeled dataset. It is challenging to train a deep neural network when there are a few training images for a specified category. In this scenario, 89 training images and six categories are introduced. However, the number of training images for direction signs, tourist signs and information signs is 3, 9 and 7, respectively. The shortage of training data has a negative impact on the supervised learning method in regard to learning the effective image features.

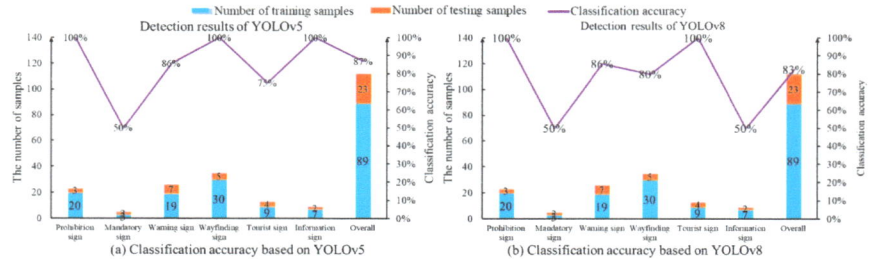

Figure 12. Classification accuracy of traffic sign images based on the YOLO programs.

Next, we apply five common quantitative evaluation metrics to check the proposed method. (1) As a straightforward method, accuracy is defined as the proportion of correct predictions to the total instances. In general, high accuracy indicates that a program has been used favorably in a detection task. However, the accuracy is not effective when the program encounters an imbalanced classification problem. (2) Precision focuses on the ratio of positive predictions created by the machine learning program which are actually correct. A high precision reveals that the program yields a low false positive rate. (3) Recall is devoted to checking the proportion of actual positive instances that have been identified. This metric is important when the cost of false negative is high. (4) As the harmonic mean of precision and recall, the F1 score attempts to find a balanced measure of the detection performance. (5) Frames Per Second (FPS) pays attention to quantifying the running speed. It represents the number of individual images can be processed by the computer program. A high FPS rate implies that the program has a high computational efficiency and real-time capability. The detailed definitions of the preceding concepts are explained in [42].

The computation results of the performance evaluation are exhibited in Table 1. The finding can be explained from twofold aspects. First, our VLM-based method has a comparable detection quality with the supervised YOLOv5. However, the YOLOv8 did not obtain a satisfactory result in this application. The values of accuracy, precision, recall and F1 score in the last row are lower than those in the first three rows. The main reason for this is that a small dataset with limited training images is not sufficient for training a YOLOv8 network. Second, the proposed method has a competitive performance in terms of running speed. Based on the VLM method, 14.98 high-resolution images can be evaluated per second. This running speed is slightly faster than the two YOLO networks. The primary factor behind the high efficiency is that CLIP focuses on understanding the global structure and extracting a visual feature vector from an input image. In comparison, the YOLO networks employ the image partition strategy. In order to explore the local pattern, the input image is uniformly divided into several patches. The high-resolution UAV images lead to an increase in the computation burden.

Table 1. Detection results of the UAV images.

Method	Accuracy (%)	Precision (%)	Recall (%)	F1 Score	FPS
VLM with dictionary1	85.6	77.8	85.0	0.80	14.98
VLM with dictionary2	89.2	75.7	78.2	0.75	14.98
YOLOv5	86.9	73.1	80.0	0.76	12.56
YOLOv8	82.6	61.9	76.0	0.68	12.53

4.4. Bootstrapping Test on the Small Dataset

In the previous section, we used a dataset with 111 drone images to examine our VLM-based method and supervised YOLO networks. However, a small dataset did not establish a solid foundation with the scientific evaluation. Therefore, we apply the bootstrapping test to check the randomness and variability with the traffic sign detection methods. The computer creates multiple new datasets by resampling the original UAV images with replacements. Image datasets with different distributions are fed into the network training procedure. The diversified datasets are helpful for checking the generalization ability of the detection method under various circumstances.

The bootstrapping test comprises three main steps. (1) A group of new training datasets are created by randomly resampling the original dataset. Each bootstrapped dataset has the same size as our UAV image dataset. A key point is that the image is sampled with replacements. This means an image can be selected multiple times in an individual bootstrapped dataset. (2) Based on the bootstrapped dataset, the training procedure is launched to upgrade the deep neural network. It should be noted that we have four programs in the previous section. First, all images in the bootstrapped dataset are used as the test material for the VLM program with keyword dictionary 1. There is no training step since the text description is generated from the Chinese national standard. Second, we utilize the BLIPv2 program to generate the keyword dictionary. For each bootstrapped dataset, 12 text descriptions are produced to depict the traffic signs in the bootstrapped dataset. The traffic sign detection task is realized by calculating the similarities between the drone images and the keyword dictionary. Third, the program trains the YOLOv5 and YOLOv8 networks. The parameter setting is the same as the previous program in Section 4.3. For each bootstrapped dataset, the first 88 images are fed into the training procedure. After the fine-tuning step, the remaining 23 images are applied as the test set within the supervised learning framework. (3) The model performance is evaluated according to the test images in the bootstrapped dataset. In this case, we apply accuracy, precision, recall and F1 score as the evaluation metrics. (4) In order to understand the robustness and stability, we combine the evaluation results from every detection program. The distribution of performance metrics becomes an important indicator of the detection quality. A small range reveals that the traffic sign prediction results are consistent across the bootstrapped samples. The detection program is not significantly influenced by the dataset variation. In comparison, a large deviation implies that there is high variability and randomness within the predictions. The detection program is sensitive to the variation within the image dataset.

Based on the preceding framework, we carry out the bootstrapping test on our drone image dataset. In particular, the number of bootstrap samples is specified as 10. This means 10 new datasets are generated from the original drone images. The distribution of evaluation metrics is shown in Figure 13. It is clear that the VLM with dictionary 1 has an impressive detection result. The detection quality is not substantially affected by the bootstrapped dataset. The main reason is that the text descriptions created using the Chinese national standard provide excellent material for expressing the intrinsic characteristics of traffic signs. In comparison, the detections of the second dictionary are not consistent across multiple bootstrapped datasets. In order to express the target instances, our program launches BLIPv2 to generate text descriptions. Given an input image, the image translation program creates a sentence to describe the main object. Thus, the quality of training images has a substantial influence on the subsequent detection results. The reliance on the specific characteristics of training images enables dictionary 2 to be sensitive to the image variation. Furthermore, a wide distribution is observed when the YOLO networks are performed in this traffic sign detection task. As the supervised learning framework, the quality and diversity are decisive factors in regard to the network convergence. An image dataset with limited training examples makes generating a robust deep neural network difficult. In some cases, the network may overfit to specific patterns or noises in the bootstrapped datasets. The preceding phenomenon reveals that it is a challenging task to train a reliable deep learning model with a small dataset.

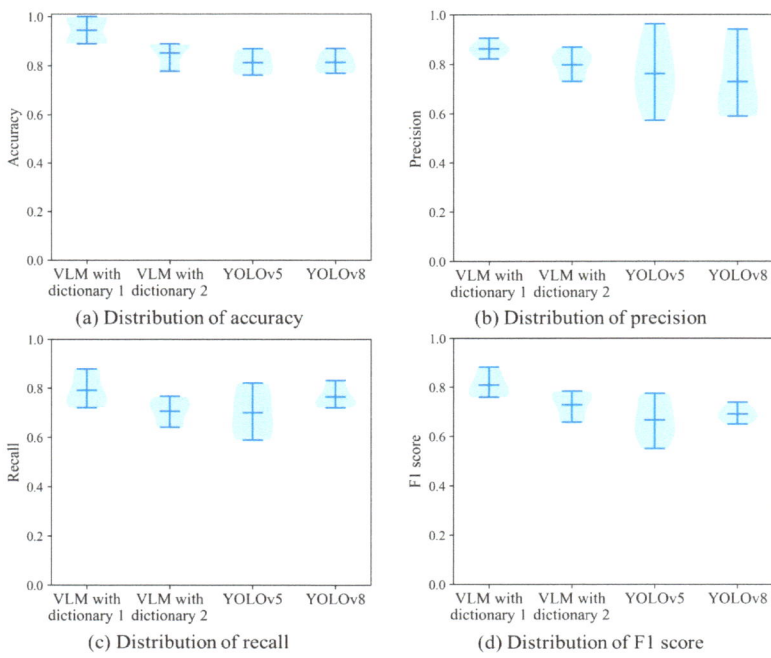

Figure 13. Distributions of four evaluation metrics in the bootstrapping test.

5. The Application of Traffic Sign Detection on Public Datasets

5.1. Experiment Configuration

In this section, two public traffic sign datasets are employed with the aim of extensively evaluating the proposed method. On one hand, we test the performance on Chinese Traffic Sign Detection Benchmark 2021 (CCTSDB2021) [42]. On the other hand, Tsinghua-Tecent 100K (TT100K) is applied to check our VLM program [43]. Since there is no fine-tuning step in our program, we focus on the test sets in CCTSDB2021 and TT100K. Three example images in these two test sets are shown in Figure 14.

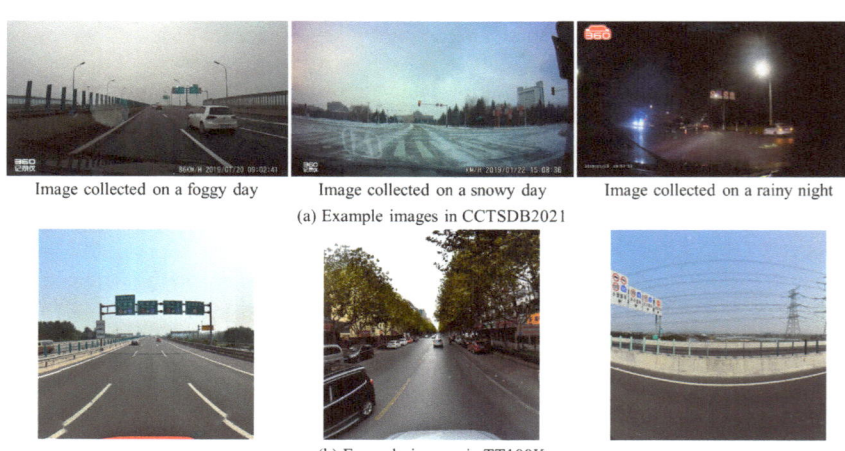

Figure 14. Example images in CCTSDB2021 and TT100K.

It is clear that there are several significant differences between our UAV dataset and the two public traffic sign datasets. (1) Complex backgrounds are involved in CCTSDB2021 and TT100K. In these two public datasets, the vehicle-based camera is used to capture the traffic sign instance. It is difficult for the detection program to distinguish the visual elements, including vehicles, traffic lights, trees and billboards. In comparison, UAV images are gathered from the aerial platform. The traffic signs are always at the center of UAV images. (2) The image resolution remarkably varies in two public datasets. Notably, 860 × 480, 1280 × 720 and 1920 × 1080 are common image sizes in CCTSDB2021. The testing image is sized 2048 × 2048 in TT100K. High-resolution images of size 8000 × 6000 are collected by the UAV. Thus, the high-resolution image provides sufficient information to realize the fine-grained analysis. (3) The images have experienced considerable visual variations in CCTSDB2021. According to the weather conditions and illumination, the testing images are partitioned into the following six groups: sunny, cloud, night, snow, foggy and rain. In contrast, there are no extreme weather conditions in the TT100K and UAV datasets. (4) The multi-object detection task is a core concept in CCTSDB2021 and TT100K. In general, there are multiple traffic sign examples in the test image. As a comparison, our UAV images concentrate on the single-object detection task. In other words, our program pays attention to identifying the main traffic sign instance within the image. (5) Three categories of traffic signs exist in CCTSDB2021 and TT100K. Prohibition, warning and mandatory sign are annotated according to the semantic meaning. On the basis of the Chinese national standard, six categories are employed in our UAV dataset.

It is worth noting that the primary reason for the difference between our UAV dataset and the two public traffic sign datasets lies in the different research problems. In our scenario, the main task is to monitor the condition of traffic signs in the context of infrastructure maintenance. The UAV technique is explored to capture images from the aerial platform. Accordingly, the traffic sign is always located at the center of UAV images. There are no extreme weather conditions or illumination issues. Despite the relatively simple visual content in UAV images, the core challenge is to fulfill the rapid deployment and realize the cross-domain adaptation. In contrast, the main task in CCTSDB2021 and TT100K is to promote the development of traffic sign detection in the field of autonomous driving. Thus, the vehicle-based camera is used to capture the traffic scenario. A competitive traffic sign detection method is supposed to deal with varying sizes, changing illumination, complex background and multiple objects.

In order to deal with CCTSDB2021 and TT100K, two modifications are conducted on our VLM-based method. (1) A pretrained YOLO network becomes a preprocessing step to localize the traffic signs in two public datasets. The program outputs a group of image patches encompassing the traffic sign instance. Based on the keyword dictionary created by the Chinese national standard, we carry out the CLIP and similarity calculations to predict the categories of investigated examples. (2) There are three specified categories of traffic signs. Our program only employs the keywords describing the prohibition, mandatory and warning signs.

5.2. Quantitative Performance Evaluation

In this section, we focus on the performance of the proposed VLM-based method on the two public traffic sign datasets. Similar to Section 4.3, a variety of evaluation metrics are used to assess the performance. Table 2 displays the detection results on two public datasets. Two key observations are found. First, the values of accuracy, precision and recall are larger than 80%. This indicates that the VLM-based program outputs reliable programs. Second, our program uses accelerating behavior to deal with two public datasets. Compared with high-resolution UAV images, the images of moderate size in CCTSDB2021 and TT100K have a positive effect on speeding up the feature extraction step.

Table 2. Detection results of our VLM-based method on two public datasets.

Dataset	Accuracy (%)	Precision (%)	Recall (%)	F1 Score	FPS
CCTSDB2021	85.9	82.5	85.4	0.83	45.18
TT100K	92.0	92.2	86.9	0.89	43.67

Next, the 1500 test images in CCTSDB2021 are further analyzed. Based on Zhang et al. [42], there are two viewpoints in regard to understanding the detection result. On one hand, the traffic sign instances are divided into five subsets according to their size. Access small (XS), small (S), medium (M), large (L) and extra large (XL) are used. On the other hand, six groups are generated on the basis of weather conditions. The extreme weather scenarios include rain, fog and snow. In addition, we introduce the detection results performed by other important detection programs. The computation results are directly cited by Zhang et al. [42,44]. Tables 3 and 4 exhibit the detection results for CCTSDB2021. In Figure 15, a column chart is created to facilitate the data visualization.

Table 3. Detection results of the CCTSDB2021 with various sign sizes.

Methods	Metrics	XS	S	M	L	XL
VLM-based method	Precision (%)	75.2	86.8	92.4	91.5	91.1
	Recall (%)	70.6	93.2	92.3	93.5	92.6
Faster RCNN [19]	Precision (%)	77.1	83.6	89.0	85.1	85.3
	Recall (%)	48.7	78.1	79.2	88.1	81.3
YOLOv5 [34]	Precision (%)	75.6	88.6	94.7	97.3	96.9
	Recall (%)	55.9	75.7	88.3	89.0	91.3
FCOS [45]	Precision (%)	85.2	91.7	93.8	91.3	93.3
	Recall (%)	35.7	60.5	83.8	93.8	90.9
One-level feature [44]	Precision (%)	77.3	85.7	89.8	90.8	88.5
	Recall (%)	86.8	82.3	90.7	97.9	99.1

Table 4. Detection results of the CCTSDB2021 with various weather conditions.

Methods	Metrics	Sunny	Cloud	Night	Rain	Foggy	Snow
VLM-based method	Precision (%)	87.1	93.5	83.7	53.9	43.1	59.3
	Recall (%)	92.1	95.9	83.8	65.1	60.8	66.7
Faster RCNN [19]	Precision (%)	85.5	92.7	76.9	61.4	77.0	96.3
	Recall (%)	77.4	57.6	47.9	34.6	67.1	91.1
YOLOv5 [34]	Precision (%)	95.9	94.0	86.1	47.9	64.8	96.1
	Recall (%)	85.1	81.2	60.6	46.7	81.3	80.7
FCOS [45]	Precision (%)	93.9	97.0	85.1	67.8	66.7	94.5
	Recall (%)	75.5	68.7	55.6	31.1	38.8	63.2
One-level feature [44]	Precision (%)	92.1	89.6	83.1	97.0	78.7	97.5
	Recall (%)	97.4	87.1	84.0	74.0	91.3	87.6

It should be noted that no fine-tuning step is performed in our program. The VLM-based method does not employ the training dataset in CCTSDB2021 and TT100K. Based on the text description created by the Chinese national standard, the computer attempts to determine the categories of traffic signs according to the distance between visual content and language features. The image variations among diverse datasets can create a substantial challenge for the detection program. In contrast, RCNN, YOLOv5, FCOS and one-level feature programs in Tables 3 and 4 belong to the supervised learning. There are two necessary steps. At first, the computer carries out the fine-tuning step to train the deep neural network. Then, the converged network is launched to identify the traffic signs in the test set. Thus, the effectiveness of the fine-tuning step has a major influence on the detection result.

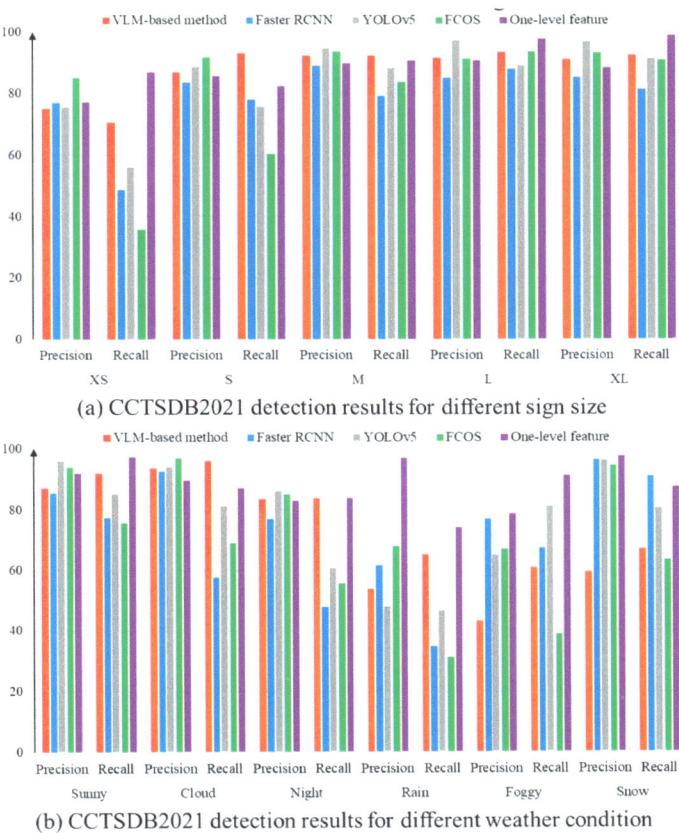

(a) CCTSDB2021 detection results for different sign size

(b) CCTSDB2021 detection results for different weather condition

Figure 15. Quantitative evaluation of the CCTSDB2021 dataset.

5.3. Discussion on the Detection Result

Based on Tables 3 and 4, there are two important findings. (1) Our VLM-based method exhibits competitive performance when the test image is taken in favorable environments. Given an input image under sunny, cloud and night conditions, our program has comparable results to the supervised detection programs. The main reason is that the generalization ability of the multi-modal learning method is helpful for dealing with the domain shift and realizing cross-domain adaptation. The visual characteristics of the traffic signs in the UAV dataset, CCTSDB2021 and TT100K are supposed to comply with Chinese regulations. Therefore, the text description provides advisable material to guide the deep learning program. The rapid deployment of traffic sign detection can be realized by exploring the consistency between the image and text description. (2) One shortcoming of the proposed method is that the detection result is highly influenced by the color information. Figure 16 displays several false detections. In our text description, the words 'red circular border' play an important role in identifying the prohibition sign. In comparison, 'blue' and 'yellow' are remarkable characteristics of mandatory and warning signs. However, the real-world case presents a difficulty. As shown in Figure 16a,c, some non-standard traffic signs are encompassed by additional information. The extra area has a perturbation on the VLM calculation. Furthermore, the small size, adverse weather conditions and image blur are negative factors of the proposed method. The absence of color information has a considerable impact on the detection result.

Figure 16. Examples of the false detections in two public datasets.

6. Conclusions

In this work, we explore a vision–language model-based traffic sign detection method to analyze high-resolution drone images. Aiming at reducing the reliance on the discrete image labels, text descriptions become important prior knowledge for guiding the image processing program. The multi-modal learning technique is launched to separately handle high-resolution images and the keyword dictionary. In addition, our program takes advantage of large-scale pretrained networks. The fine-tuning step is saved to realize the rapid deployment. There are three main components. First, a keyword dictionary is built to provide text descriptions. On one hand, we refer to the regulations in the Chinese national standard. On the other hand, the BLIPv2 model is activated to perform image translation. Second, we perform the multi-modal learning to achieve the feature extraction. Based on the encoder network within the CLIP framework, UAV images and text descriptions are independently converted into the high-dimensional feature vectors. Third, the similarity between images and keywords becomes an applicable tool for determining the categories of traffic signs. Cosine distance and the Softmax function are carried out to compare the visual features and text embeddings.

A practical application in Guyuan, China, is used to test the proposed method. A number of 111 UAV high-resolution images are captured to monitor the condition of traffic sign instances. Further examinations are performed on two public datasets, CCTSDB2021 and TT100K. Based on the experiment results, there are two primary advantages to our VLM-based method. (1) The dependency on the human annotations is significantly alleviated by the employment of multi-modal learning. Rather than discrete labels, the text description becomes a feasible means of explaining the characteristics of traffic signs. According to the computation results, the keyword dictionary is beneficial for addressing the domain shift among diverse datasets. (2) With the objective of realizing rapid deployment, there is no fine-tuning step in our program. The pretrained VLM is used to analyze the input images as well as text descriptions. The experiment results indicate that our program has a strong generalization ability and cross-domain adaptation potential.

As a preliminary exploration of a vision–language model in the context of traffic sign detection, our program contains two important limitations. First, the detection quality is heavily dependent on the color information. The low visibility and image blur have a negative impact on the calculation result. Second, our program classifies the traffic sign instances into six categories. A small number of categories does not convey abundant information or provide substantial guidance for the intelligent transportation system. Therefore, a potential extension of this work could help to improve detection accuracy under adverse weather conditions. It is feasible to apply the prompt engineering technique to create high-quality text descriptions. Furthermore, another future research direction is to introduce cutting-edge multi-modal techniques. A combination of self-supervised learning and traffic sign datasets is an applicable method for understanding task-specific knowledge.

Author Contributions: Conceptualization, J.Y. and J.L.; methodology, J.Y., J.L. and Y.L.; software, Y.L. and M.Z.; validation, C.Z., S.D. and Z.D.; writing—original draft preparation, J.Y. and J.L.; writing—review and editing, C.Z., S.D. and Z.D.; funding acquisition, J.Y. and J.L. All authors have read and agreed to the published version of the manuscript.

Funding: This research was funded by the Chinese Ministry of Transportation In Service Trunk Highway Infrastructure and Safety Emergency Digitization Project (No. 2023-26) and Transportation Research Project of Department of Transport of Shaanxi Province (No. 23-04X).

Data Availability Statement: The data presented in this study are available on request from the corresponding author due to the privacy restriction.

Conflicts of Interest: Authors Jianqun Yao, Jinming Li and Yuxuan Li are employed by the company CCCC Infrastructure Maintenance Group Co., Ltd. The remaining authors declare that the research was conducted in the absence of any commercial or financial relationships that could be construed as a potential conflict of interest.

References

1. Canese, L.; Cardarilli, G.C.; Di Nunzio, L.; Fazzolari, R.; Famil Ghadakchi, H.; Re, M.; Spanò, S. Sensing and Detection of Traffic Signs Using CNNs: An Assessment on Their Performance. *Sensors* **2022**, *22*, 8830. [CrossRef] [PubMed]
2. Sanyal, B.; Mohapatra, R.K.; Dash, R. Traffic Sign Recognition: A Survey. In Proceedings of the 2020 International Conference on Artificial Intelligence and Signal Processing (AISP), Amaravati, India, 10–12 January 2020.
3. Lim, X.R.; Lee, C.P.; Lim, K.M.; Ong, T.S.; Alqahtani, A.; Ali, M. Recent Advances in Traffic Sign Recognition: Approaches and Datasets. *Sensors* **2023**, *23*, 4674. [CrossRef] [PubMed]
4. Chakravarthy, A.S.; Sinha, S.; Narang, P.; Mandal, M.; Chamola, V.; Yu, F.R. DroneSegNet: Robust Aerial Semantic Segmentation for UAV-Based IoT Applications. *IEEE Trans. Veh. Technol.* **2022**, *71*, 4277–4286. [CrossRef]
5. Arya, D.; Maeda, H.; Sekimoto, Y. From Global Challenges to Local Solutions: A Review of Cross-country Collaborations and Winning Strategies in Road Damage Detection. *Adv. Eng. Inform.* **2024**, *60*, 102388. [CrossRef]
6. Du, J.; Zhang, R.; Gao, R.; Nan, L.; Bao, Y. RSDNet: A New Multiscale Rail Surface Defect Detection Model. *Sensors* **2024**, *24*, 3579. [CrossRef]
7. Madani, A.; Yusof, R. Traffic sign recognition based on color, shape, and pictogram classification using support vector machines. *Neural Comput. Appl.* **2018**, *30*, 2807–2817. [CrossRef]
8. Kerim, A.; Efe, M.Ö. Recognition of Traffic Signs with Artificial Neural Networks: A Novel Dataset and Algorithm. In Proceedings of the 2021 International Conference on Artificial Intelligence in Information and Communication (ICAIIC), Jeju Island, Republic of Korea, 13–16 April 2021; pp. 171–176.
9. Soni, D.; Chaurasiya, R.K.; Agrawal, S. Improving the Classification Accuracy of Accurate Traffic Sign Detection and Recognition System Using HOG and LBP Features and PCA-Based Dimension Reduction. In Proceedings of the International Conference on Sustainable Computing in Science, Technology and Management (SUSCOM), Amity University Rajasthan, Jaipur, India, 20–22 January 2019.
10. Namyang, N.; Phimoltares, S. Thai traffic sign classification and recognition system based on histogram of gradients, color layout descriptor, and normalized correlation coefficient. In Proceedings of the 2020-5th International Conference on Information Technology (InCIT), Chonburi, Thailand, 21–22 October 2020; pp. 270–275.
11. Wang, B. Research on the Optimal Machine Learning Classifier for Traffic Signs. In Proceedings of the SHS Web of Conferences, Virtual Event, 26 August 2022; EDP Sciences: Les Ulis, France, 2022; Volume 144, p. 03014.
12. Li, J.; Wang, Z. Real-time traffic sign recognition based on efficient CNNs in the wild. *IEEE Trans. Intell. Transp. Syst.* **2018**, *20*, 975–984. [CrossRef]

13. Sokipriala, J.; Orike, S. Traffic sign classification comparison between various convolution neural network models. *Int. J. Sci. Eng. Res.* **2021**, *12*, 165–171. [CrossRef]
14. Zhu, Y.; Yan, W.Q. Traffic sign recognition based on deep learning. *Multimed. Tools Appl.* **2022**, *81*, 17779–17791. [CrossRef]
15. Redmon, J.; Divvala, S.; Girshick, R.; Farhadi, A. You Only Look Once: Unified, Real-Time Object Detection. In Proceedings of the 2016 IEEE Conference on Computer Vision and Pattern Recognition (CVPR), Las Vegas, NV, USA, 27–30 June 2016; pp. 779–788.
16. Li, X.; Geng, S. Improved traffic sign detection for YOLOv5s. In Proceedings of the IEEE 4th International Conference on Computer Engineering and Application, Hangzhou, China, 7–8 April 2023; pp. 696–699.
17. Yu, J.; Ye, X.; Tu, Q. Traffic Sign Detection and Recognition in Multiimages Using a Fusion Model With YOLO and VGG Network. *Trans. Intell. Transp. Syst.* **2022**, *23*, 16632–16642. [CrossRef]
18. Girshick, R.; Donahue, J.; Darrell, T.; Malik, J. Region-Based Convolutional Networks for Accurate Object Detection and Segmentation. *IEEE Trans. Pattern Anal. Mach. Intell.* **2015**, *38*, 142–158. [CrossRef]
19. Ren, S.; He, K.; Girshick, R.; Sun, J. Faster R-CNN: Towards real-time object detection with region proposal networks. *Trans. Pattern Anal. Mach. Intell.* **2017**, *39*, 1137–1149. [CrossRef] [PubMed]
20. Zhang, J.; Xie, Z.; Sun, J.; Zou, X.; Wang, J. A cascaded R-CNN with multiscale attention and imbalanced samples for traffic sign detection. *IEEE Access* **2020**, *8*, 29742–29754. [CrossRef]
21. Zhang, J.; Wang, W.; Lu, C.; Wang, J.; Sangaiah, A.K. Lightweight deep network for traffic sign classification. *Ann. Telecommun.* **2020**, *75*, 369–379. [CrossRef]
22. Triki, N.; Karray, M.; Ksantini, M. A Real-Time Traffic Sign Recognition Method Using a New Attention-Based Deep Convolutional Neural Network for Smart Vehicles. *Appl. Sci.* **2023**, *13*, 4793. [CrossRef]
23. Zhang, J.; Ye, Z.; Jin, X.; Wang, J.; Zhang, J. Real-time traffic sign detection based on multiscale attention and spatial information aggregator. *J. Real-Time Image Process.* **2022**, *19*, 1155–1167. [CrossRef]
24. Carion, N.; Massa, F.; Synnaeve, G.; Usunier, N.; Kirillov, A.; Zagoruyko, S. End-to-end object detection with transformers. In Proceedings of the European Conference on Computer Vision, Glasgow, UK, 23–28 August 2020; pp. 213–229.
25. Zhang, J.; Huang, J.; Jin, S.; Lu, S. Vision-Language Models for Vision Tasks: A Survey. *arXiv* **2023**, arXiv:2304.00685. [CrossRef]
26. Jaiswal, A.; Ramesh Babu, A.; Zaki Zadeh, M.; Banerjee, D.; Makedon, F. A Survey on Contrastive Self-Supervised Learning. *Technologies* **2021**, *9*, 2. [CrossRef]
27. Gui, J.; Chen, T.; Zhang, J.; Cao, Q.; Sun, Z.; Luo, H.; Tao, D. A Survey on Self-supervised Learning: Algorithms, Applications, and Future Trends. *arXiv* **2023**, arXiv:2301.05712. [CrossRef]
28. Khan, S.; Naseer, M.; Khan, S.; Naseer, M.; City, M.; Dhabi, A.; Zamir, S.W.; Shah, M.; Hayat, M.; Zamir, S.W.; et al. Transformers in Vision: A Survey. *ACM Comput. Surv. (CSUR)* **2022**, *54*, 200. [CrossRef]
29. Dosovitskiy, A.; Beyer, L.; Kolesnikov, A.; Weissenborn, D.; Zhai, X.; Unterthiner, T.; Dehghani, M.; Minderer, M.; Heigold, G.; Gelly, S.; et al. An Image is Worth 16×16 Words: Transformers for Image Recognition at Scale. *arXiv* **2021**, arXiv:2010.11929.
30. Liu, Z.; Lin, Y.; Cao, Y.; Hu, H.; Wei, Y.; Zhang, Z.; Lin, S.; Guo, B. Swin Transformer: Hierarchical Vision Transformer using Shifted Windows. *arXiv* **2021**, arXiv:2103.14030.
31. Redmon, J.; Farhadi, A. YOLO9000: Better, faster, stronger. In Proceedings of the 2017 IEEE Conference on Computer Vision and Pattern Recognition (CVPR), Honolulu, HI, USA, 21–26 July 2017; pp. 7263–7271.
32. Redmon, J.; Farhadi, A. Yolov3: An incremental improvement. *arXiv* **2018**, arXiv:1804.02767.
33. Bochkovskiy, A.; Wang, C.-Y.; Liao, H.-Y.M. Yolov4: Optimal speed and accuracy of object detection. *arXiv* **2020**, arXiv:2004.10934.
34. Ultralytics. YOLOv5. Available online: https://github.com/ultralytics/yolov5 (accessed on 1 November 2020).
35. Wang, C.-Y.; Bochkovskiy, A.; Liao, H.-Y.M. YOLOv7: Trainable bag-of-freebies sets new state-of-the-art for real-time object detectors. In Proceedings of the 2023 IEEE/CVF Conference on Computer Vision and Pattern Recognition (CVPR), Vancouver, BC, Canada, 17–24 June 2023; pp. 7464–7475.
36. Jocher, G.; Chaurasia, A.; Qiu, J. Ultralytics YOLOv8. Available online: https://github.com/ultralytics/ultralytics (accessed on 20 June 2024).
37. *GB 5768-2022*; Traffic Signs. Standardization Administration of China: Beijing, China, 2022.
38. Li, J.; Li, D.; Xiong, C.; Hoi, S. BLIP: Bootstrapping Language-Image Pre-training for Unified Vision-Language Understanding and Generation. *arXiv* **2022**, arXiv:2201.12086v2.
39. Li, J.; Li, D.; Savarese, S.; Hoi, S. BLIP-2: Bootstrapping Language-Image Pre-Training with Frozen Image Encoders and Large Language Models. *arXiv* **2023**, arXiv:2301.12597.
40. Radford, A.; Kim, J.W.; Hallacy, C.; Ramesh, A.; Goh, G.; Agarwal, S.; Sastry, G.; Askell, A.; Mishkin, P.; Clark, J.; et al. Learning transferable visual models from natural language supervision. In Proceedings of the 38th International Conference on Machine Learning, PMLR, Virtual Event, 18–24 July 2021; pp. 8748–8763.
41. Vaswani, A.; Shazeer, N.; Parmar, N.; Uszkoreit, J.; Jones, L.; Gomez, A.N.; Kaiser, Ł.; Polosukhin, I. Attention Is All You Need. *arXiv* **2017**, arXiv:1706.03762.
42. Zhang, J.; Zou, X.; Kuang, L.D.; Wang, J.; Sherratt, R.S.; Yu, X. CCTSDB 2021: A more comprehensive traffic sign detection benchmark. *Hum.-Centric Comput. Inf. Sci.* **2022**, *12*, 23.
43. Zhu, Z.; Liang, D.; Zhang, S.; Huang, X.; Li, B.; Hu, S. Traffic-sign detection and classification in the wild. In Proceedings of the IEEE Conference on Computer Vision and Pattern Recognition, Las Vegas, NV, USA, 26 June–1 July 2016; pp. 2110–2118.

44. Zhang, J.; Lv, Y.; Tao, J.; Huang, F.; Zhang, J. A robust real-time anchor-free traffic sign detector with one-level feature. *IEEE Trans. Emerg. Top. Comput. Intell.* **2024**, *8*, 1437–1451. [CrossRef]
45. Tian, Z.; Shen, C.; Chen, H.; He, T. FCOS: A simple and strong anchor free object detector. *IEEE Trans. Pattern Anal. Mach. Intell.* **2022**, *44*, 1922–1933. [CrossRef]

Disclaimer/Publisher's Note: The statements, opinions and data contained in all publications are solely those of the individual author(s) and contributor(s) and not of MDPI and/or the editor(s). MDPI and/or the editor(s) disclaim responsibility for any injury to people or property resulting from any ideas, methods, instructions or products referred to in the content.

Article

Visual-Inertial RGB-D SLAM with Encoder Integration of ORB Triangulation and Depth Measurement Uncertainties

Zhan-Wu Ma and Wan-Sheng Cheng *

School of Electronic and Information Engineering, University of Science and Technology Liaoning, Anshan 114051, China
* Correspondence: cws@ustl.edu.cn

Abstract: In recent years, the accuracy of visual SLAM (Simultaneous Localization and Mapping) technology has seen significant improvements, making it a prominent area of research. However, within the current RGB-D SLAM systems, the estimation of 3D positions of feature points primarily relies on direct measurements from RGB-D depth cameras, which inherently contain measurement errors. Moreover, the potential of triangulation-based estimation for ORB (Oriented FAST and Rotated BRIEF) feature points remains underutilized. To address the singularity of measurement data, this paper proposes the integration of the ORB features, triangulation uncertainty estimation and depth measurements uncertainty estimation, for 3D positions of feature points. This integration is achieved using a CI (Covariance Intersection) filter, referred to as the CI-TEDM (Triangulation Estimates and Depth Measurements) method. Vision-based SLAM systems face significant challenges, particularly in environments, such as long straight corridors, weakly textured scenes, or during rapid motion, where tracking failures are common. To enhance the stability of visual SLAM, this paper introduces an improved CI-TEDM method by incorporating wheel encoder data. The mathematical model of the encoder is proposed, and detailed derivations of the encoder pre-integration model and error model are provided. Building on these improvements, we propose a novel tightly coupled visual-inertial RGB-D SLAM with encoder integration of ORB triangulation and depth measurement uncertainties. Validation on open-source datasets and real-world environments demonstrates that the proposed improvements significantly enhance the robustness of real-time state estimation and localization accuracy for intelligent vehicles in challenging environments.

Keywords: covariance intersection filter; encoders; RGB-D SLAM; multisensor fusion

Citation: Ma, Z.-W.; Cheng, W.-S. Visual-Inertial RGB-D SLAM with Encoder Integration of ORB Triangulation and Depth Measurement Uncertainties. *Sensors* **2024**, *24*, 5964. https://doi.org/10.3390/s24185964

Academic Editors: Chunhui Zhao and Shuai Hao

Received: 6 August 2024
Revised: 5 September 2024
Accepted: 12 September 2024
Published: 14 September 2024

Copyright: © 2024 by the authors. Licensee MDPI, Basel, Switzerland. This article is an open access article distributed under the terms and conditions of the Creative Commons Attribution (CC BY) license (https://creativecommons.org/licenses/by/4.0/).

1. Introduction

In recent years, with the rapid development of robotics, computer vision, autonomous driving, augmented reality/virtual reality, planetary exploration, and unmanned aerial vehicle navigation [1], visual odometry and SLAM have emerged as prominent areas of research [2]. SLAM technology enables robots to not only achieve real-time self-localization but also to construct and continuously update maps of unknown environments during exploration, significantly enhancing their ability to operate autonomously in such settings [3]. In the field of robotics, intelligent wheeled robots have emerged as a prominent area of research [4]. Based on the types of sensors employed, mainstream SLAM systems are primarily classified into laser-based SLAM, vision-based SLAM, and various sensor-assisted laser/vision multisensor fusion SLAM technologies. Visual SLAM relies on cameras as the primary sensors, and its ability to capture rich information, combined with its lightweight and cost-effective nature, has garnered significant interest from researchers [5]. Within the domain of visual SLAM, ORB-SLAM3 demonstrates superior performance, surpassing other algorithms in terms of accuracy, computational efficiency, and robustness [6]. Currently, ORB-SLAM3 acquires the 3D positions of feature points by directly utilizing RGB-D depth camera measurements, without accounting for the uncertainty in triangulation estimation. Furthermore, ORB-SLAM3 fuses data from the RGB-D depth camera and the IMU, but it does not integrate wheel encoder data, which are crucial for improving SLAM accuracy and aiding mobile robot localization and navigation. The literature [7] proposes a lightweight multisensor fusion method involving ORB-SLAM, IMU, and wheel odometry for the localization and navigation of an indoor mobile robot in a GPS-denied environment. Experimental results demonstrate that the robot can localize itself with tolerable error and exhibit strong navigation capabilities in specific scenarios. The ORB-SLAM in this study employs a monocular camera to compute the real-time camera position through feature matching. To filter out dynamic objects during localization and mapping, the literature [8] presents a mobile robot localization method using asynchronous data fusion comprising a wheel odometer and a visual odometer within an Extended Kalman Filter (EKF) framework, based on a semantic SLAM system. The approach integrates semantic SLAM based on ORB-SLAM2 with YOLOv3. The asynchronous data fusion for mobile robot localization, comprising visual and wheel odometer data, is implemented at a frequency of approximately 50 Hz using EKF. Experimental results demonstrate that this method effectively improves the accuracy and robustness of the mobile robot. Recently, the literature [9] proposed a visual-inertial-wheel odometry method that provides robust initialization and highly accurate estimates for ground robots. This study utilizes a novel maximum-a-posteriori initialization, coupled with wheel encoder measurements, to address the unobservable scale problem in visual-inertial-only initialization during straight-line motion at the beginning of the trajectory. Experiments on public datasets demonstrate the effectiveness and efficiency of the initialization, the robustness of pose tracking, and the improved accuracy of the entire trajectory. In addition, the literature [10] utilizes a CI filter to fuse inertial navigation sensors and wheel odometry to improve the accuracy of real-time localization and mapping. The effectiveness and feasibility of the algorithm are validated using an experimental platform. The aforementioned literature demonstrates that the use of wheel encoders not only enhances the accuracy and robustness of SLAM but also addresses the initialization failures in mobile robots. Inspired by this literature, this paper proposes a novel tightly coupled visual-inertial RGB-D SLAM with encoder integration of ORB triangulation and depth measurement uncertainties to improve ORB-SLAM3.

Visual odometry (VO) algorithms can be categorized into two main types: feature-based methods and direct methods [11]. However, in both feature-based and direct methods within SLAM, accurately estimating the uncertainty of the camera pose and the 3D positions of feature points is crucial. Such accurate estimations are essential for selecting precise keyframes, particularly in the context of information fusion and active SLAM [12,13]. The primary objective is to minimize uncertainty or entropy, which necessitates a closed-form solution for the uncertainty estimation of the camera pose and the 3D positions of feature points. The uncertainty estimation of the camera pose [1] and the 3D positions of feature points are key parameters for SLAM. This uncertainty can generally be expressed using the covariance matrix [14], information entropy [15], and Fisher's information matrix [13]. The estimations of uncertainty in the camera poses and the 3D positions of feature points are primarily based on filters (e.g., the Extended Kalman Filter) [16] and nonlinear optimization methods (e.g., Bundle Adjustment and graph optimization) [17]. Uncertainty estimations based on filter methods are both convenient and fast, providing direct access to the covariance matrix or expected entropy. In contrast, nonlinear optimization-based methods do not directly estimate uncertainty. As a result, the covariance matrix and information entropy are not optimized as direct parameters. Instead, uncertainty in nonlinear optimization requires rigorous computation [18].

In the study of Vakhitov et al. [19], a PnP(L) solver based on EPnP and DLS is proposed for uncertainty-aware pose estimation by considering both the 3D coordinates and 2D projections of feature points in SLAM. Additionally, the motion-only bundle adjustment is modified to account for the uncertainty in the 3D positions of feature points. Tests performed on the KITTI datasets demonstrate improved accuracy. In Belter et al. (2016) [20], two spatial uncertainty models are proposed, based on experimental tests, to estimate the covariance matrix of the measured feature points. The resulting covariance is then incorporated into factor graph optimization in the back-end of the SLAM system. While these methods provide error models and uncertainties for 3D points, they cannot be directly applied to more general visual SLAM systems. In Belter et al. (2018) [21], the problem of estimating the uncertainty in the 3D positions of feature points using the feature method is discussed in detail. The effect of point feature uncertainty on trajectory and map uncertainty estimation is investigated through an uncertainty model. A factor graph optimization model is employed to minimize the error in the 3D positions of feature points measured by depth cameras. Experiments conducted on open-source datasets demonstrate that the algorithm can improve the accuracy of camera pose estimation.

Currently, researchers mainly focus on one type of uncertainty study, such as the 3D positions of feature points [22–24] or camera poses [25–28]. Li and Yang [29] present a pose fusion method that accounts for the possible correlations among measurements. The handling of these correlations is based on the theory of Covariance Intersection (CI), where the independent and dependent parts are separated to yield a more consistent result. Few studies have investigated the problem of joint uncertainty estimation for the 3D positions of feature points and camera pose. However, the above methods are unable to provide closed-form solutions for uncertainty estimations, which are crucial for information fusion and active SLAM, as motion prediction and pose evaluation require one or even multiple steps of forward-looking uncertainty propagation. To address the above problems, this paper proposes a closed-form uncertainty estimation algorithm, called CI-TEDM (Covariance Intersection for Triangulation Estimates and Depth Measurements), for fusing uncertainties in triangulation estimates and depth measurements of the 3D positions of feature points. This algorithm is based on the closed-form uncertainty estimation of both camera poses and 3D positions of feature points. On one hand, based on the ORB-SLAM3 system, the covariance matrix of the camera pose and the triangulation of ORB features is estimated separately. On the other hand, the covariance matrix of the 3D positions of feature points obtained from the depth measurements of the RGB-D camera is computed. The 3D positions of feature points estimated by the two approaches are then fused with a Covariance Intersection (CI) filter, which uses the two covariance matrices as weights to achieve optimal fusion [18].

Moreover, the initialization of monocular SLAM can be time-consuming when the system lacks substantial disparity. Binocular SLAM requires more time to process stereo pairs [30]. Therefore, in this paper, RGB-D SLAM is used to achieve robustness and real-time performance. The open-source ORB-SLAM3 system with monocular, stereo, and RGB-D cameras has attracted much attention from scholars for its high accuracy and real-time performance [6]. Inspired by the IMU pre-integration model and the powerful and fast loop closure detection in [31], this paper introduces wheel encoder edges, which optimize the state of each frame to avoid tracking failures in RGB-D cameras and help improve the robustness of the entire RGB-D SLAM system [32]. Although RGB-D cameras can achieve centimeter-level or even higher accuracy, vision SLAM still faces significant challenges in complex environments, such as changes in illumination, moving objects, adverse weather, weak textures, and environmental degradation [33]. Cameras are prone to losing tracking information, which can lead to tracking failures due to fast camera motion, unstable performance of embedded boards, and low frame rates during image capture [34]. In this study, auxiliary sensors, such as IMU and encoders, are integrated to mitigate the problem of temporary tracking failures that may occur in pure-vision SLAM. Visual-inertial SLAM is a current research hotspot and can be categorized into two types: loosely coupled and tightly coupled [35]. Tightly coupled SLAM systems include direct methods [36], keyframe-based visual-inertial SLAM (OKVIS) [37], robust visual-inertial odometry based on EKF [38], and the monocular visual-inertial system (VINS-Mono) [39]. Among these methods, the most accurate results for the EuRoC datasets [40] are achieved by visual-inertial ORB-SLAM3, which is primarily based on the theoretical framework of real-time IMU pre-integration [6]. In this study, encoder and IMU data are fully utilized to enhance the robustness and real-time performance of RGB-D ORB-SLAM3.

The main contribution of this paper lies in the new combination of CI-TEDM and wheel odometry, which seeks to improve the accuracy and robustness of SLAM. Additionally, this paper derives the wheel encoder pre-integration model, error model, rotation-to-state-variable Jacobian matrix, and position-to-state-variable Jacobian matrix, which are used in local BA optimization to enhance SLAM results.

The rest of this paper is organized as follows. Section 2 outlines the framework of the system. Section 3 describes the fusion of ORB-SLAM3 triangulation estimation and depth measurement estimation algorithms. Section 4 derives the encoder pre-integration model, which provides the Jacobian matrix of the rotation-to-state variable and the position-to-state variable to facilitate local BA optimization for improved SLAM results. Section 5 validates the algorithms using open-source datasets and real-world environments, presenting the experimental results. Section 6 concludes the paper.

2. System Overview

The system framework diagram is shown in Figure 1, which consists of the input, function and output of the three main modules. The primary enhancement in the input module is the addition of a new encoder data thread. The system inputs include ORB images and RGB-D images from the RGB-D camera, encoder data from the wheel encoder, and IMU data from the IMU. In the function module, ORB feature points are first extracted and matched from each RGB image. The uncertainty in the camera pose is then derived using these matched point pairs. Given the observation noise of feature points in the RGB image plane, the uncertainty in the 6D camera pose ξ, represented by an element of the Lie algebra se(3), is estimated through implicit differentiation and covariance propagation. Utilizing the resulting camera pose uncertainty, the uncertainty in the triangulation of ORB feature points is computed, leading to the estimation of the uncertainty in the position of each 3D point. Simultaneously, the uncertainty in the depth measurements from the RGB-D camera is propagated to determine the uncertainty in the position of each 3D point. The uncertainties from both triangulation and depth measurements are then fused using a CI filter. Additionally, it is required to synchronize the encoder and IMU data with image frames within a permissible time alignment error. Encoder or IMU data that exceed the time alignment threshold relative to the nearest keyframe are discarded. The encoder and IMU data are then pre-integrated to compute the wheel and inertial odometry between adjacent image frames. The system with CI-TEDM that fuses IMU and encoder data based on ORB-SLAM3 is referred to as VIEOS3-TEDM, the system fusing only encoder data is termed VEOS3-TEDM, and the system using CI-TEDM alone is called VOS3-TEDM. An important modification is the addition of pure encoder edges in VEOS3-TEDM, which connect two states, x_i and x_j, using only encoder measurements and are incorporated into the spanning tree. This tree primarily propagates corrected poses to all keyframes in the map. Finally, the CI filter fusion results and odometry results are input into the tracking, local mapping, loop closure, map merging, and full bundle adjustment (BA) modules for processing. In the output module, the map points of the environment and the keyframes of the mobile robot's trajectory are obtained through the aforementioned processing.

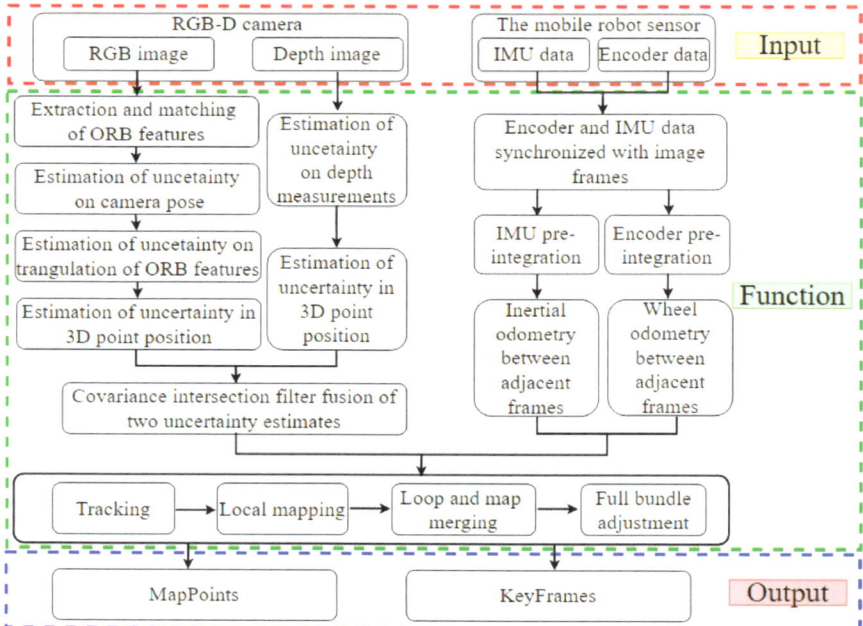

Figure 1. The system framework diagram. The system framework diagram consists of three main modules: input, function, and output.

3. Fusion of ORB-SLAM3 Triangulation and Depth Measurement Uncertainty Estimations

3.1. Uncertainty Estimation in ORB-SLAM3 Triangulation and Depth Measurement

The first step is to estimate the uncertainty of the camera pose. Minimizing the reprojection error of 3D feature points is commonly used to estimate the camera pose, as illustrated in Figure 2. Once the uncertainty of the camera pose is determined, the uncertainty of the 3D feature points can be calculated. Simultaneously, the depth measurement uncertainty is also estimated. The derivation of the solution formula mentioned above can be found in Yuan et al. [18].

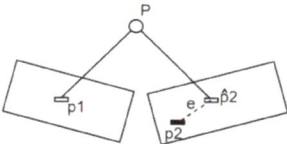

Figure 2. An example diagram of reprojection error. Feature matching indicates that points $p1$ and $p2$ are projections of the same spatial point p, but the camera pose is initially unknown. Initially, there is a certain distance between the projected point, $\hat{p}2$, of P and the actual point, $p2$. The camera pose is then adjusted to minimize this distance.

3.2. Fusion of Two Uncertainty Estimations

It is assumed that the two uncertainty estimates of the 3D feature points obtained using the camera follow a normal distribution, as denoted in Yuan et al. [18].

$$N\left({}^w\hat{p}_{i,Tr}, \text{cov}\left({}^w\hat{p}_{i,Tr}\right)\right) \text{ and } N\left({}^w\hat{p}_{i,Dep}, \text{cov}\left({}^w\hat{p}_{i,Dep}\right)\right),$$

where $^w\hat{p}_{i,Tr}$ and $\text{cov}\left(^w\hat{p}_{i,Tr}\right)$ represent the uncertainty concerning the world coordinate system (denoted as w) and the covariance matrix of the 3D feature points triangulated by the camera, respectively. These estimates are then fused to obtain a high-precision map. According to the derivation formula for triangulation and depth measurement uncertainty estimation, it is evident that a correlation exists between the two normal distributions. In this paper, a CI filter is employed to fuse these two correlated estimates, which allows for the fusion of two estimates with unknown cross-covariance and ensures consistent results. The formula [18] is as follows:

$$\text{cov}\left(^w\hat{p}_i\right) = \left[\omega \text{cov}^{-1}\left(^w\hat{p}_{i,Tr}\right) + (1-\omega) \cdot \text{cov}^{-1}\left(^w\hat{p}_{i,Dep}\right)\right]^{-1} \quad (1)$$

$$^w\hat{p}_i = \text{cov}\left(^w\hat{p}_i\right) \cdot \left[\omega \text{cov}^{-1}\left(^w\hat{p}_{i,Tr}\right)^w \cdot \hat{p}_{i,Tr} + (1-\omega) \cdot \text{cov}^{-1}\left(^w\hat{p}_{i,Dep}\right)^w \cdot \hat{p}_{i,Dep}\right] \quad (2)$$

where $\omega \in [0,1]$ is the weight assigned to the uncertainty in the triangulation estimate and the depth measurement estimate, which can be calculated by minimizing the trace of the fused covariance [18], i.e.,

$$\omega^* = \underset{\omega}{\text{argmin}}\left\{tr\left[\text{cov}\left(^w\hat{p}_i\right)\right]\right\}. \quad (3)$$

4. Derivation of the Wheel Encoder Model

4.1. Pre-Integration Model for Wheeled Encoder

There are multiple wheel encoder measurements between two adjacent frames [8]. In this paper, inspired by IMU pre-integration [6], a wheel encoder pre-integration model is derived to compute a low-frequency integration term that provides motion constraints for the two frames. Since the object of study is a ground mobile robot, its motion is approximated to be in a plane. Thus, the motion of the robot can be represented by a rotation angle, $\theta \in R$, around the z-axis and a 2D translation vector, $q \in R^2$, in the $x - y$ plane. The wheel encoder provides the mobile robot with measured values for linear velocity, v, and angular velocity, ω [41]. It is assumed that the rear wheel of the mobile robot is aligned with the positive x-axis of the wheel encoder coordinate system and that the left side of the rear wheel is aligned with the positive y-axis. The wheel encoder coordinate system follows the right-hand rule, and the movement model for the wheel odometry is shown in Figure 3. Figure 3 shows the robot moving from coordinate (x_k, y_k) at time t_k to the coordinate (x_{k+1}, y_{k+1}) at time t_{k+1}. The linear velocity of the mobile robot is measured by an incremental photoelectric encoder, which converts pulse data into the robot's linear velocity equation:

$$v = \frac{\pi d_w}{N \Delta t} n, \quad (4)$$

where d_w is the diameter of the rear wheel of the robot, N is the number of encoders lines, Δt is the sampling time, and n is the number of pulses during the sampling time. The linear velocities of the robot's drive wheels are calculated by Equation (4), and the true velocities of the left and right wheels are obtained by subtracting measurement noise from the observed values as follows [42]:

$$v_k^{el} = \tilde{v}_k^{el} - \eta_k^{eld}, \quad (5)$$

$$v_k^{er} = \tilde{v}_k^{er} - \eta_k^{erd}, \quad (6)$$

where $\begin{bmatrix} \eta_k^{eld} & \eta_k^{erd} \end{bmatrix}^T \in \mathbb{R}^2$ is the measurement noise caused by the wheel encoders, which is assumed to follow a zero-mean Gaussian distribution with covariance matrix $\begin{bmatrix} \eta_k^{eld} & \eta_k^{erd} \end{bmatrix}^T \in \mathbb{R}^2$.

v_k^{el} and v_k^{er} are the estimated instantaneous linear velocities at moment t_k. $\widetilde{v}_k^{el}$ and $\widetilde{v}_k^{er}$ are the linear velocities converted from the measurements of the left and right wheel encoders at moment t_k. It is assumed that the robot operates in an ideal environment where the surface is flat and the wheels do not slip. The linear velocity of the wheels is denoted by $v_k^e = [\widetilde{v}_k^{ex} \ 0 \ 0]^T - \eta_k^{mvd}$, where $\eta_k^{m\omega d} = [\eta_k^{m\omega x} \ \eta_k^{m\omega y} \ \eta_k^{m\omega z}]^T \in \mathbb{R}^3$ denotes the measurement noise vector in the direction of x, y, z for the linear velocity. The angular velocity of the wheels is denoted by $\omega_k^e = [0 \ 0 \ \widetilde{\omega}_k^{ez}]^T - \eta_k^{m\omega d}$, where $\widetilde{\omega}_k^{ez}$ is the angular velocity converted from the measured value of the wheel encoder at moment t_k. $\eta_k^{m\omega d} = [\eta_k^{m\omega x} \ \eta_k^{m\omega y} \ \eta_k^{m\omega z}]^T \in \mathbb{R}^3$ represents the measurement noise vector in the direction of x, y, z for the angular velocity.

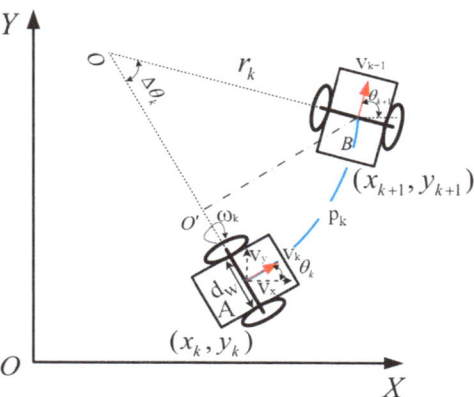

Figure 3. The motion model of the wheeled robot using wheel encoders. The figure illustrates the motion model of a mobile robot using wheel encoders in a 2D plane. The model describes the robot's trajectory between its position at time t_k, denoted as (x_{k+1}, y_{k+1}), and its position at time t_{k+1}, denoted as (x_{k+1}, y_{k+1}).

Due to the high frequency of the wheel encoder, the state transition model between two adjacent frames of wheel encoder measurements can be described as:

$$\theta_{k+1} = \theta_k + \omega_k \cdot \Delta t, \tag{7}$$

$$q_{k+1} = q_k + R(\theta_k) \cdot \begin{bmatrix} 0 \\ v_k \cdot \Delta t \end{bmatrix}. \tag{8}$$

Here, the Euler integration method is used, and by calculating Equations (7) and (8) in recursive form, the wheel encoder pre-integration term between two adjacent frames can be obtained. Since the ground robot operates in a 2D $x - y$ plane, the rotation matrix is expressed as:

$$R(\theta_k) = \begin{bmatrix} \cos\theta_k & -\sin\theta_k \\ \sin\theta_k & \cos\theta_k \end{bmatrix}. \tag{9}$$

Due to noise in the wheel encoder measurements, the following Equation (10) is used to construct the measurement error model:

$$\begin{aligned} \overline{\omega} &= \omega + n_\omega \\ \overline{v}_x &= v_x + n_x \\ \overline{v}_y &= v_y + n_y \end{aligned} . \tag{10}$$

In Equation (10), v_x and v_y denote the forward and left lateral linear velocities of the mobile robot, respectively. Assuming the noise of ω, v_x, and v_y is Gaussian white noise, with Gaussian distributions as follows: $n_\omega \sim N(0, \delta_\omega^2)$, $n_{v_x} \sim N(0, \delta_{v_x}^2)$, $n_{v_y} \sim N(0, \delta_{v_y}^2)$.

A linear recurrence relation for the state error of the pre-integrated term of the wheel encoder can be derived from Equations (7), (8) and (10):

$$\begin{bmatrix} \delta\theta_{k+1} \\ \delta q_{k+1} \end{bmatrix} = P_k \begin{bmatrix} \delta\theta_k \\ \delta q_k \end{bmatrix} + H_k \begin{bmatrix} n_\omega \\ n_v \end{bmatrix}. \tag{11}$$

In Equation (11), n_v denotes the measurement noise of v_x and v_y. The Jacobian matrices P_k and H_k can be calculated from Equations (12) and (13), respectively:

$$P_k = \begin{bmatrix} 1 & 0_{1\times 2} \\ R(\theta_k) \cdot J \cdot \begin{bmatrix} 0 \\ v_k \cdot \Delta t \end{bmatrix} & I_{2\times 2} \end{bmatrix}, \tag{12}$$

$$H_k = \begin{bmatrix} \Delta t & 0_{1\times 2} \\ 0_{2\times 1} & R(\theta_k) \cdot \Delta t \end{bmatrix}, \tag{13}$$

$$J = \begin{bmatrix} 0 & -1 \\ 1 & 0 \end{bmatrix}. \tag{14}$$

After obtaining the linear transition equation for the state error at adjacent moments, the covariance matrix of the wheeled encoder pre-integration term between neighboring image frames can be calculated using the following Equation (15):

$$Q_{k+1} = P_k Q_k P_k^T + H_k \sum_n H_k^T, \tag{15}$$

where Q_k is the covariance matrix of the state quantities θ_k and q_k; $\sum_n$ is the covariance matrix of the noise, which is calculated by the following Equation (16):

$$\sum_n = \begin{bmatrix} \delta_\omega^2 & 0 & 0 \\ 0 & \delta_{v_x}^2 & 0 \\ 0 & 0 & \delta_{v_y}^2 \end{bmatrix}. \tag{16}$$

4.2. Pre-Integration Error of Wheeled Encoder

Based on the derivation of wheel encoder pre-integration in Section 4.1, the pre-integration error for the wheel encoders is [42]:

$$r_{\Delta\phi_{ij}^e}^e \triangleq Log[(\Delta \widetilde{R}_{ij}^e)^T \Delta R_{ij}^e], \tag{17}$$

$$r_{\Delta p_{ij}^e}^e \triangleq \Delta p_{ij}^e - \Delta \widetilde{p}_{ij}^e. \tag{18}$$

In the above equation, ΔR_{ij}^e and Δp_{ij}^e are the predicted values of the wheel encoders pre-integration, and $\Delta \widetilde{R}_{ij}^e$ and $\Delta \widetilde{p}_{ij}^e$ are the observed values of the wheel encoders pre-integration.

4.2.1. Jacobian Matrix of Rotation to State Variables

Equation (17) represents the rotational error in the pre-integration of the wheel encoders, which does not involve p_{wi}^e and p_{wj}^e. Therefore, the Jacobian matrix with respect to each of these state variables is 0, and its Jacobian matrix with respect to $\delta\phi_i^e$ and $\delta\phi_j^e$ is given by [42]:

$$\frac{\partial r_{\Delta\phi_{ij}^e}^e}{\partial \delta\phi_i^e} = -J_r^{-1}(r_{\Delta\phi_{ij}^e}^e)(R_{wi}^{eT} R_{wj}^e R_{ce})^T, \tag{19}$$

$$\frac{\partial r^e_{\Delta\phi^e_{ij}}}{\partial \delta\phi^e_j} = J_r^{-1}(r^e_{\Delta\phi^e_{ij}})R^T_{ce}. \tag{20}$$

In the above equation, R_{ce} is the rotation matrix from the wheel encoder coordinate system to the camera coordinate system, which can be obtained from the intelligent vehicle model built in ROS.

4.2.2. Jacobian Matrix of Position to State Variables

Equation (18) represents the pre-integrated position error of the wheel encoders. Its Jacobian matrix with respect to p^e_{wi}, p^e_{wj}, $\delta\phi_i$ and $\delta\phi_j$ is given by [42]:

$$\frac{\partial r^e_{\Delta p^e_{ij}}}{\partial \delta p^e_{wi}} = -R^T_{ce}R^{eT}_{wi}, \tag{21}$$

$$\frac{\partial r^e_{\Delta p^e_{ij}}}{\partial \delta p^e_{wj}} = R^{eT}_{ce}R^{eT}_{wi}, \tag{22}$$

$$\frac{\partial r^e_{\Delta p^e_{ij}}}{\partial \delta\phi^e_i} = R^T_{ce}(R^{eT}_{wi}(R^e_{wj}p_{ce} + p^e_{wj} - p^e_{wi}))^{\wedge}, \tag{23}$$

$$\frac{\partial r^e_{\Delta p^e_{ij}}}{\partial \delta\phi^e_j} = -R^T_{ce}R^{eT}_{wi}R^e_{wj}p^{\wedge}_{ce}. \tag{24}$$

In the above equation, p_{ce} is the translation matrix from the wheel encoder coordinate system to the camera coordinate system, which can be obtained from the intelligent vehicle model built in ROS.

5. Experimental Analysis of the VEOS3-TEDM Algorithm

5.1. Experimental Analysis of Open-Source Datasets

The combination of the three datasets—RGB-D, IMU, and wheel encoders—is not currently available in some public datasets. To validate the effectiveness of the VEOS3-TEDM algorithm proposed in this paper, open-source datasets from the literature [42] are used. These datasets are applied to a differential wheeled robot equipped with a Kinect v2 RGB-D camera, two-wheel encoders, and an IMU sensor. The ground truth data for these datasets are provided via a total station, which generates the true trajectory of the robot at approximately 10 Hz by measuring the position of a prismatic reflector fixed to the robot, with an accuracy close to 1 mm. These datasets were recorded using a mobile robot on an experimental platform, with captured scenes including long straight corridors and laboratories. To validate the deviation of the estimated trajectories from the true trajectories, a public benchmarking tool [43] was used to evaluate the Absolute Trajectory Error (ATE), which reflects the SLAM accuracy by calculating the Root-Mean-Squared Error (RMSE) of the system output. For evaluating algorithm accuracy, only the translation error is considered, and thus the Average Translation Error is defined as follows:

$$ATE_{trans} = \sqrt{\frac{1}{N}\sum_{i=1}^{N}\left\|T_{trans}(T^{-1}_{gt,i}T_{esti,i})\right\|^2_2}. \tag{25}$$

In the above equation, $T_{gt,i}$ and $T_{esti,i}$ represent the true and estimated values, respectively. T_{trans} denotes the translation portion of the variable inside the parentheses. The RMSE values in Tables 1 and 2 were obtained using an EVO (Evaluation of Odometry, version 1.29.0) tool. The VEOS3-TEDM algorithm was evaluated using two datasets: one from a long straight corridor scene and another from a laboratory scene. The frame rate of the camera was 15 Hz, while the IMU and wheel encoders operated at 200 Hz. The corridor dataset contains 897 frames, and the laboratory dataset contains 875 frames. Table 1 shows the results of the algorithm comparison for the corridor dataset, while Table 2 presents the results for the laboratory dataset. An "X" in the table indicates that the value could not be computed, meaning that the algorithm failed to track or complete initialization. The pose estimation ratio is defined as the number of frames successfully tracked by the algorithm divided by the total number of frames in the dataset.

Table 1. The comparison of algorithm results on the corridor dataset.

Algorithm Name	Sensors	RMSE (m)	Average Tracking Time per Frame (ms)	Pose Estimation Ratio (%)
	Encoders	0.463	1	100
ORB-SLAM2	RGB-D	X	X	45
ORB-SLAM3	RGB-D	X	X	94
	RGB-D + IMU	X	X	5
VEORB-SLAM3	RGB-D + Encoders	0.094	17	100
VIEORB-SLAM3	RGB-D + IMU + Encoder	s0.114	26	100
VOS3-TEDM	CI-TEDM	X	X	96
VEOS3-TEDM	CI-TEDM + Encoders	0.083	14	100
VIEOS3-TEDM	CI-TEDM + IMU + Encoders	0.107	23	100

Table 2. The comparison of algorithm results on the laboratory dataset.

Algorithm Name	Sensors	RMSE (m)	Average Tracking Time per Frame (ms)	Pose Estimation Ratio (%)
	Encoders	0.382	1	100
ORB-SLAM2	RGB-D	X	X	92
ORB-SLAM3	RGB-D	0.1167	22	100
	RGB-D + IMU	X	X	5
VEORB-SLAM3	RGB-D + Encoders	0.1104	21	100
VIEORB-SLAM3	RGB-D + IMU + Encoders	0.128	33	100
VOS3-TEDM	CI-TEDM	0.1054	20	100
VEOS3-TEDM	CI-TEDM + Encoders	0.087	16	100
VIEOS3-TEDM	CI-TEDM + IMU + Encoders	0.115	29	100

The texture of the corridor in those datasets is very weak. During the turn of the intelligent vehicle, the camera is very close to the white wall surface, resulting in a very small number of feature points or even none at all. This causes purely visual SLAM tracking to fail, so both the ORB-SLAM2 and ORB-SLAM3 algorithms using the RGB-D camera fail to track. In Table 1, it can be observed that two types of algorithms are unable to compute the RMSE and the average tracking time per frame. However, ORB-SLAM3, with its multiple map modes, achieves a final pose estimation ratio that is twice as high as that of the ORB-SLAM2 algorithm. When ORB-SLAM3 utilizes the combined form of RGB-D and IMU, the IMU pre-initialization requires the intelligent vehicle to perform sufficient motion to achieve the necessary excitation for IMU initialization. During this time, visual tracking must not fail for an extended period. Nevertheless, visual tracking tends to fail during the vehicle's turning process. The IMU was unable to complete initialization within the dataset. While the intelligent vehicle's pose estimation can be achieved using only the wheel encoder, this method results in the shortest average tracking time per frame but produces a relatively high RMSE, which does not meet practical requirements.

As shown in Table 1, the VEOS3-TEDM algorithm proposed in this paper outperforms the other algorithms in the corridor datasets. Specifically, in terms of RMSE, it achieves reductions of 38 cm, 1.1 cm, 3.1 cm, and 2.4 cm compared to the encoders and VEORB-SLAM3, VIEORB-SLAM3, and VIEOS3-TEDM algorithms, respectively. Regarding average tracking time per frame, the VEOS3-TEDM algorithm is 3 ms, 12 ms, and 9 ms faster than the VEORB-SLAM3, VIEORB-SLAM3, and VIEOS3-TEDM algorithms, respectively. These results indicate that the VEOS3-TEDM algorithm offers superior performance in both RMSE and average tracking time per frame within the corridor dataset.

The laboratory dataset is rich in texture information, but the high-speed motion of the intelligent vehicle causes tracking failures in ORB-SLAM2. After adding the IMU, ORB-SLAM3 has more stringent initialization requirements, which the intelligent vehicle platform struggles to meet. This results in the inability to calculate the RMSE and average tracking time per frame. However, the final pose estimation ratios of ORB-SLAM3 and ORB-SLAM2 are nearly identical in these datasets. Table 2 shows that, in the laboratory dataset, the VEOS3-TEDM algorithm proposed in this paper achieves RMSE reductions of 29.5 cm, 2.97 cm, 2.34 cm, 4.1 cm, 1.84 cm, and 2.8 cm compared to the encoders and ORB-SLAM3, VEORB-SLAM3, VIEORB-SLAM3, VOS3-TEDM, and VIEOS3-TEDM algorithms, respectively. In terms of average tracking time per frame, VEOS3-TEDM is 6 ms, 5 ms, 17 ms, 4 ms, and 13 ms faster than ORB-SLAM3, VEORB-SLAM3, VIEORB-SLAM3, VOS3-TEDM, and VIEOS3-TEDM, respectively. These results imply that the VEOS3-TEDM algorithm outperforms other algorithms in both RMSE and average tracking time per frame in the laboratory dataset.

Tables 1 and 2 show that the ORB-SLAM3 algorithm fails to track both in the corridor and laboratory datasets when implemented using RGB-D and IMU fusion. This failure is attributed to the lack of feature points, fast motion, and light reflections. However, the VEOS3-TEDM algorithm proposed in this paper effectively addresses these temporary failure issues, achieving satisfactory accuracy and successfully recovering the entire trajectory. The results demonstrate that the VEOS3-TEDM algorithm generally achieves better accuracy and stability due to the fusion of ORB triangulation estimation with RGB-D depth measurement and the integration of encoder information. This approach aids in identifying outliers, enhancing prediction, and improving pose estimation. The improvements are particularly noticeable in datasets with higher motion speeds.

The process of running the VEOS3-TEDM algorithm on the corridor and laboratory datasets is illustrated in Figure 4. During the execution, the pose of all keyframes is based on the world coordinate system, which is established at the frame where initialization is completed. After initialization, map points are created and their 3D coordinates are relative to the world coordinate system. The map points are categorized into two types: local map points and global map points. Local map points assist in localization during the tracking process. In Figure 4, the blue frames represent keyframes, red frames represent initial keyframes, and green frames represent current frames. The VEOS3-TEDM algorithm achieves satisfactory accuracy and successfully recovers the entire trajectory. This demonstrates that the VEOS3-TEDM algorithm, which integrates ORB triangulation estimation, RGB-D depth measurement, and encoder information, effectively fulfills the primary goal of this paper.

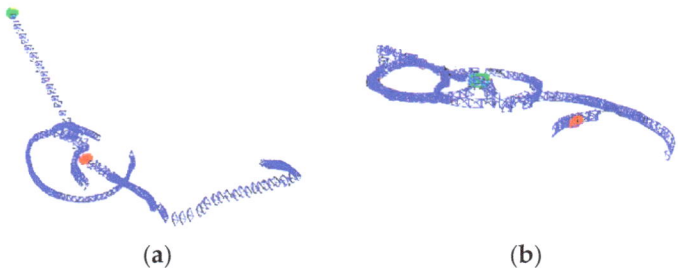

(a) (b)

Figure 4. The process of running datasets in the VEOS3-TEDM algorithm: (**a**) corridor scene and (**b**) laboratory scene. The blue frames represent keyframes, red frames represent initial keyframes, and green frames represent current frames.

The tracking process of the VEOS3-TEDM algorithm on the corridor and laboratory datasets is shown in Figure 5, where the green boxes indicate the recognized 3D feature points. In Figure 5, it is evident that angular feature points and those with significant changes in gray value are well identified. This demonstrates that the VEOS3-TEDM algorithm can stably track the image without failure. This stability is achieved by incorporating encoder information into the algorithm. Even if the visual tracking fails, the encoder data can compensate, allowing the algorithm to connect the two states and maintain continuous stable tracking.

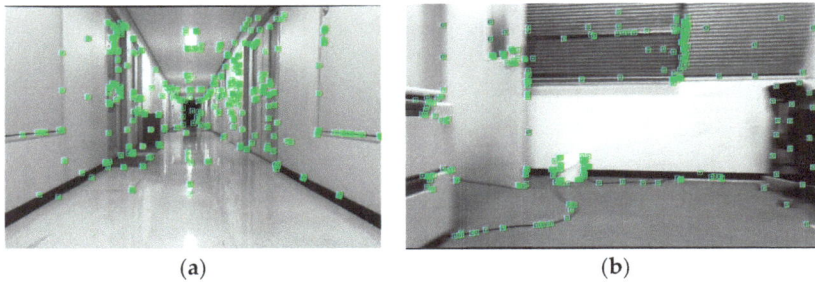

(a) (b)

Figure 5. The process of tracking datasets in the VEOS3-TEDM algorithm: (**a**) corridor scene and (**b**) laboratory scene. The green boxes in the figure represent key feature points detected by VEOS3-TEDM algorithm.

When the VEOS3-TEDM algorithm finishes running, it generates the pose information of the keyframes. Figure 6 shows a comparison between the pose estimates generated by the VEOS3-TEDM algorithm using the open-source software EVO and the true trajectories. As seen in Figure 6, the estimated trajectories closely match the true trajectories on flat and straight roads, with only minor errors observed in the corridor scene. However, there is some deviation in the vehicle's trajectory at turns, where the camera is very close to the white wall, resulting in few or no feature points. This causes the pure vision ORB-SLAM3 tracking to fail. Nonetheless, the VEOS3-TEDM algorithm is able to track continuously and stably due to the compensation provided by encoder data. In the laboratory scene, the estimated trajectory of the vehicle closely matches the true trajectory, as the scene is rich in texture information, enabling accurate recognition of 3D feature points, and thus producing a more accurate estimated trajectory.

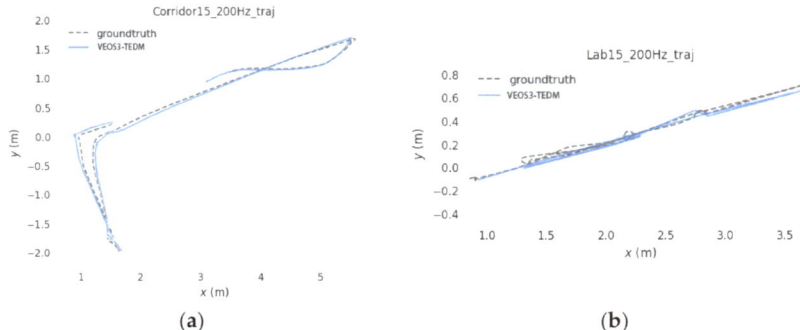

Figure 6. The comparison between estimated and true trajectory in the VEOS3-TEDM algorithm: (**a**) corridor scene and (**b**) laboratory scene.

As shown in Figure 7, the trajectory estimation errors in the x-axis and y-axis directions are relatively small in both the corridor and laboratory scenes. However, the errors in the z-axis direction are comparatively larger. This is due to the fact that encoder information is recorded more accurately in the horizontal direction, while larger errors are observed in the vertical direction. This suggests that further improvements are needed in the algorithms to enhance the accuracy of depth value processing. Additionally, the trajectory estimation errors in the x-axis and y-axis directions are smaller in the corridor scene compared to the laboratory scene. This is likely because the corridor scene is more open, with fewer obstacles, leading to a more accurate trajectory estimation in the horizontal directions.

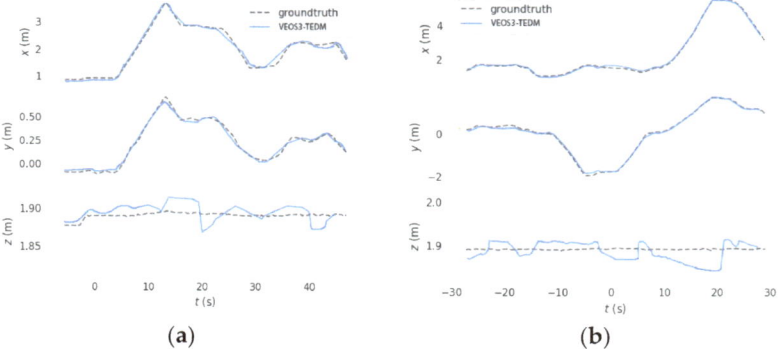

Figure 7. The comparison between true and estimated trajectories in x, y and z directions, using the VEOS3-TEDM algorithm: (**a**) corridor scene and (**b**) laboratory scene.

Finally, the intelligent vehicle successfully constructed 3D point cloud maps of the corridor and laboratory scenes, as shown in Figure 8. The constructed 3D point cloud maps are consistent with the actual scenes. For approximate navigation, the positional accuracy of obstacles, which are defined as points located between the ground plane and the height of the robot, is considered acceptable. However, some redundancies remain due to inevitable errors introduced by the pure encoders, which can be minimized through more closed-loop detection, image measurements, and better prior calibration. The point cloud representation of the map points is mainly used to highlight visible errors in the estimated trajectories of the keyframes. However, for ROS navigation, future research should focus on using a grid map form that incorporates advanced filtering techniques [43].

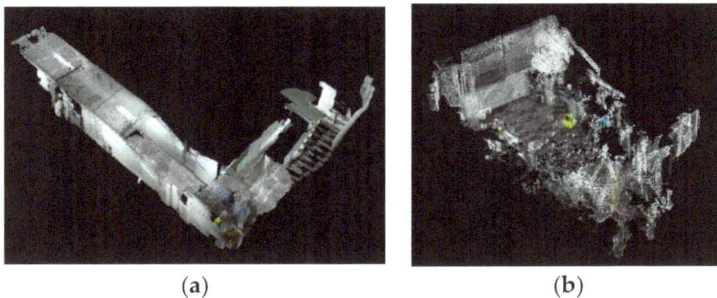

Figure 8. 3D point cloud maps: (**a**) corridor scene and (**b**) laboratory scene.

5.2. Experimental Analysis of Real-World Environments

The experimental platform is shown in Figure 9. The mobile robot is divided into a bottom layer, an intermediate layer, and an upper layer. The bottom layer contains the lithium battery, lower computer, motor driver, software-based emergency stop button, servo motor, and motor. The middle layer houses the single-wire LiDAR, power switch, and emergency stop button. The upper layer contains the upper computer, RGB-D camera, mechanical arm, voltage reduction module, and power switch for the upper computer. The locations of the devices held in the bottom and upper layers of the mobile robot are shown in Figure 10.

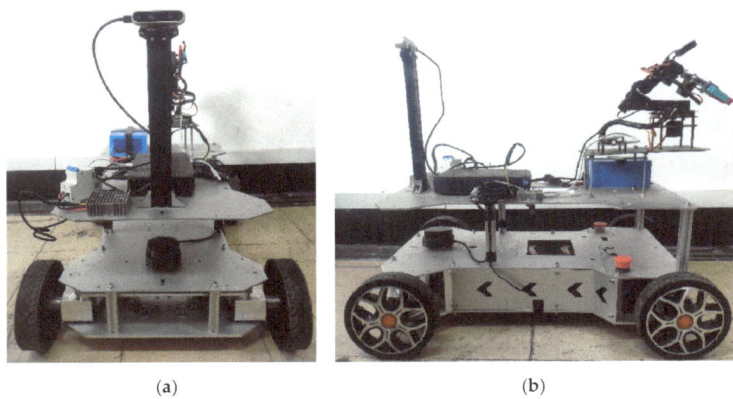

Figure 9. Images of the experimental platform: (**a**) front view and (**b**) left view.

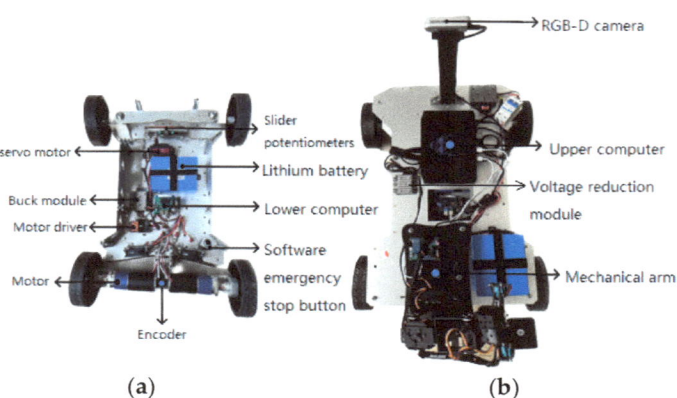

Figure 10. The location of various components on the mobile robot: (**a**) bottom level and (**b**) upper level.

The proposed VEOS3-TEDM algorithm is first ported to a mobile robot. Afterward, the algorithm is validated using the mobile robot. In this paper, data sequences of real-world indoor scenes are captured by an experimental platform. For the collected indoor scene data, it is challenging to obtain the ground truth for each moment on the experimental platform. Instead, the dataset is collected by controlling the motion of the experimental platform so that the data sequence forms a large closed loop at the end. Specifically, the experimental platform starts at the initial point, completes a full circle, and returns to the starting point. Thus, the start and end segments of the data sequence are in the same scenario.

The RGB-D camera used in the process has a frame rate of 30 Hz and an image resolution of 640 × 480. The IMU operates at 200 Hz, and the wheel encoders operate at 100 Hz. The data from the wheel encoders are collected by the lower computer and transmitted to the upper computer via serial communication. The operating environments include the laboratory, a long straight corridor, and a hall. These environments are relatively large, and the tracking process of the experimental platform during the experiment is shown in Figure 11. Figure 11 indicates that the experimental platform captures a relatively large number of 3D feature points in the laboratory, corridor, and hall scenes, including distinct angular points and points with significant changes in gray value. This demonstrates that the VEOS3-TEDM algorithm achieves the desired accuracy and stability. However, the 3D feature points obtained in weakly textured scenes (e.g., white walls) are relatively sparse, suggesting that texture richness directly impacts the accuracy and stability of localization and map building. Therefore, to achieve high localization and map-building accuracy, the mobile robot should operate in texture-rich scenes.

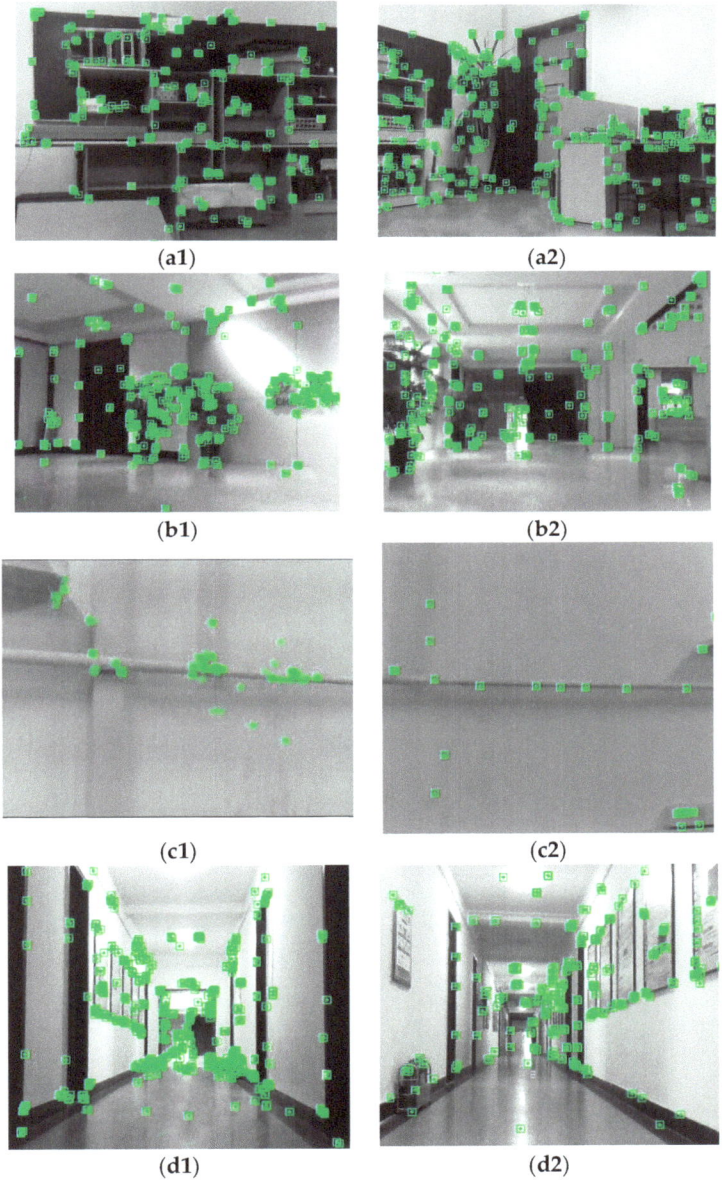

Figure 11. The process of tracking real-world environments in the VEOS3-TEDM algorithm: (**a1**,**a2**) laboratory, (**b1**,**b2**) hall, (**c1**,**c2**) weak texture scene, (**d1**,**d2**) long straight corridor. The green boxes in the figure represent key feature points detected by VEOS3-TEDM algorithm.

Figure 12 illustrates the estimated trajectory of VEOS3-TEDM running in a real-world environment compared to the ground truth, with an obtained RMSE of 0.091. The effectiveness and robustness of the proposed VEOS3-TEDM algorithm were validated in real-world environments. Additionally, the estimated trajectory closely aligns with the ground truth, achieving the research objectives outlined in this paper.

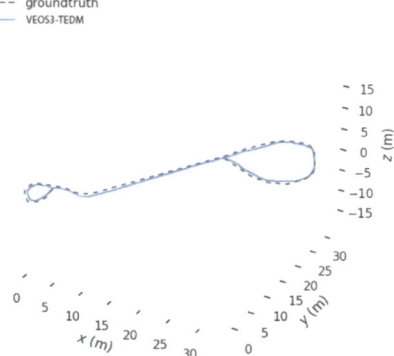

Figure 12. A comparison of estimated and true trajectories in real-world environments using the VEOS3-TEDM algorithm.

6. Conclusions

This paper introduces wheel odometry into the classical ORB-SLAM3 framework. The main contribution is the proposal of a novel combination of CI-TEDM and wheel odometry, offering a new approach to enhancing the accuracy and robustness of SLAM. Additionally, this paper derives the wheel encoder pre-integration model, error model, rotation-to-state-variable Jacobian matrix, and position-to-state-variable Jacobian matrix, all of which are employed for local BA optimization to improve SLAM results. Through these advancements, a more accurate and robust VEOS3-TEDM algorithm is achieved. Experiments on open-source datasets and in real-world environments demonstrate that the proposed VEOS3-TEDM algorithm surpasses state-of-the-art methods in both positioning and mapping accuracy. Furthermore, the proposed method is extensible to other optimization-based indirect visual SLAM or visual-inertial SLAM systems.

Author Contributions: Conceptualization, Z.-W.M. and W.-S.C.; methodology, Z.-W.M. and W.-S.C.; software, Z.-W.M.; validation, Z.-W.M.; formal analysis, Z.-W.M. and W.-S.C.; resources, Z.-W.M. and W.-S.C.; data curation, Z.-W.M.; writing—original draft preparation, Z.-W.M.; writing—review and editing, Z.-W.M. and W.-S.C.; visualization, Z.-W.M. All authors have read and agreed to the published version of the manuscript.

Funding: This research received no external funding.

Institutional Review Board Statement: Not applicable.

Informed Consent Statement: Not applicable.

Data Availability Statement: Data are contained within the article.

Conflicts of Interest: The authors declare no conflicts of interest.

References

1. Zhao, C.H.; Fan, B.; Hu, J.; Pan, Q.; Xu, Z. Homography-based camera pose estimation with known gravity direction for UAV navigation. *Sci. China Inf. Sci.* **2021**, *64*, 1–3. [CrossRef]
2. Wang, K.; Zhao, G.; Lu, J. A Deep Analysis of Visual SLAM Methods for Highly Automated and Autonomous Vehicles in Complex Urban Environment. *IEEE Trans. Intell. Transp. Syst.* **2024**, *4*, 1–18. [CrossRef]
3. Zhao, Y.L.; Hong, Y.T.; Huang, H.P. Comprehensive Performance Evaluation between Visual SLAM and LiDAR SLAM for Mobile Robots: Theories and Experiments. *Appl. Sci.* **2024**, *14*, 3945. [CrossRef]
4. Hu, X.; Zhu, L.; Wang, P.; Yang, H.; Li, X. Improved ORB-SLAM2 mobile robot vision algorithm based on multiple feature fusion. *IEEE Access* **2023**, *11*, 100659–100671. [CrossRef]
5. Al-Tawil, B.; Hempel, T.; Abdelrahman, A.; Al-Hamadi, A. A review of visual SLAM for robotics: Evolution, properties, and future applications. *Front. Robot. AI* **2024**, *11*, 1347985. [CrossRef]

6. Campos, C.; Elvira, R.; Rodríguez, J.J.; Montiel, J.M.; Tardós, J.D. Orb-slam3: An accurate open-source library for visual, visual–inertial, and multimap slam. *IEEE Trans. Robot.* **2021**, *37*, 1874–1890. [CrossRef]
7. Lin, J.; Peng, J.; Hu, Z.; Xie, X.; Peng, R. Orb-slam, imu and wheel odometry fusion for indoor mobile robot localization and navigation. *Acad. J. Comput. Inf. Sci* **2020**, *27*, 131–141.
8. Lee, C.; Peng, J.; Xiong, Z. Asynchronous fusion of visual and wheel odometer for SLAM applications. In Proceedings of the 2020 IEEE/ASME International Conference on Advanced Intelligent Mechatronics (AIM), Boston, MA, USA, 6–9 July 2020; pp. 1990–1995.
9. Zhou, W.; Pan, Y.; Liu, J.; Wang, T.; Zha, H. Visual-Inertial-Wheel Odometry With Wheel-Aided Maximum-a-Posteriori Initialization for Ground Robots. *IEEE Robot. Autom. Lett.* **2024**, *9*, 4814–4821. [CrossRef]
10. Anousaki, G.; Gikas, V.; Kyriakopoulos, K. INS-Aided Odometry and Laser Scanning Data Integration for Real Time Positioning and Map-Building of Skid-Steered Vehicles. In Proceedings of the International Symposium on Mobile Mapping Technology, Padua, Italy, 29–31 May 2007.
11. Zhou, W.; Zhou, R. Vision SLAM algorithm for wheeled robots integrating multiple sensors. *PLoS ONE* **2024**, *19*, e0301189. [CrossRef]
12. Cabrera-Ávila, E.V.; da Silva, B.M.; Gonçalves, L.M. Nonlinearly Optimized Dual Stereo Visual Odometry Fusion. *J. Intell. Robot. Syst.* **2024**, *110*, 56. [CrossRef]
13. Chen, Y.; Huang, S.; Fitch, R. Active SLAM for mobile robots with area coverage and obstacle avoidance. *IEEE/ASME Trans. Mechatron.* **2020**, *25*, 1182–1192. [CrossRef]
14. Chen, Y.; Huang, S.; Zhao, L.; Dissanayake, G. Cramér–Rao bounds and optimal design metrics for pose-graph SLAM. *IEEE Trans. Robot.* **2021**, *37*, 627–641. [CrossRef]
15. Mu, B.; Giamou, M.; Paull, L.; Agha-mohammadi, A.A.; Leonard, J.; How, J. Information-based active SLAM via topological feature graphs. In Proceedings of the 2016 IEEE 55th Conference on Decision and Control (CDC), Las Vegas, NV, USA, 12–14 December 2016; pp. 5583–5590.
16. Li, N.; Zhou, F.; Yao, K.; Hu, X.; Wang, R. Multisensor Fusion SLAM Research Based on Improved RBPF-SLAM Algorithm. *J. Sens.* **2023**, *2023*, 3100646. [CrossRef]
17. Lin, X.; Huang, Y.; Sun, D.; Lin, T.Y.; Englot, B.; Eustice, R.M.; Ghaffari, M. A Robust Keyframe-Based Visual SLAM for RGB-D Cameras in Challenging Scenarios. *IEEE Access* **2023**, *11*, 97239–97249. [CrossRef]
18. Yuan, J.; Zhu, S.; Tang, K.; Sun, Q. ORB-TEDM: An RGB-D SLAM approach fusing ORB triangulation estimates and depth measurements. *IEEE Trans. Instrum. Meas.* **2022**, *71*, 1–15. [CrossRef]
19. Vakhitov, A.; Ferraz, L.; Agudo, A.; Moreno-Noguer, F. Uncertainty-aware camera pose estimation from points and lines. In Proceedings of the IEEE/CVF Conference on Computer Vision and Pattern Recognition, Nashville, TN, USA, 20–25 June 2021; pp. 4659–4668.
20. Belter, D.; Nowicki, M.; Skrzypczyński, P. Improving accuracy of feature-based RGB-D SLAM by modeling spatial uncertainty of point features. In Proceedings of the 2016 IEEE International Conference on Robotics and Automation (ICRA), Stockholm, Sweden, 16–21 May 2016; pp. 1279–1284.
21. Belter, D.; Nowicki, M.; Skrzypczyński, P. Modeling spatial uncertainty of point features in feature-based RGB-D SLAM. *Mach. Vis. Appl.* **2018**, *29*, 827–844. [CrossRef]
22. Zhang, L.; Zhang, Y. Improved feature point extraction method of ORB-SLAM2 dense map. *Assem. Autom.* **2022**, *42*, 552–566. [CrossRef]
23. Lee, S.W.; Hsu, C.M.; Lee, M.C.; Fu, Y.T.; Atas, F.; Tsai, A. Fast point cloud feature extraction for real-time slam. In Proceedings of the 2019 International Automatic Control Conference (CACS), Keelung, Taiwan, 13–16 November 2019; pp. 1–6.
24. Zhou, F.; Zhang, L.; Deng, C.; Fan, X. Improved point-line feature based visual SLAM method for complex environments. *Sensors* **2021**, *21*, 4604. [CrossRef]
25. Gomez-Ojeda, R.; Moreno, F.A.; Zuniga-Noël, D.; Scaramuzza, D.; Gonzalez-Jimenez, J. PL-SLAM: A stereo SLAM system through the combination of points and line segments. *IEEE Trans. Robot.* **2019**, *35*, 734–746. [CrossRef]
26. Engel, J.; Schöps, T.; Cremers, D. LSD-SLAM: Large-scale direct monocular SLAM. In Proceedings of the European Conference on Computer Vision, Zurich, Switzerland, 6–12 September 2014; Volume 8690, pp. 834–849.
27. Eudes, A.; Lhuillier, M. Error propagations for local bundle adjustment. In Proceedings of the 2009 IEEE Conference on Computer Vision and Pattern Recognition, Miami, FL, USA, 20–25 June 2009; pp. 2411–2418.
28. Barfoot, T.D.; Furgale, P.T. Associating uncertainty with three-dimensional poses for use in estimation problems. *IEEE Trans. Robot.* **2014**, *30*, 679–693. [CrossRef]
29. Li, L.; Yang, M. Joint localization based on split covariance intersection on the Lie group. *IEEE Trans. Robot.* **2021**, *37*, 1508–1524. [CrossRef]
30. Mur-Artal, R.; Montiel, J.M.; Tardos, J.D. ORB-SLAM: A versatile and accurate monocular SLAM system. *IEEE Trans. Robot.* **2015**, *31*, 1147–1163. [CrossRef]
31. Wang, Y.; Ng, Y.; Sa, I.; Parra, A.; Rodriguez, C.; Lin, T.J.; Li, H. Mavis: Multi-camera augmented visual-inertial slam using se2 (3) based exact imu pre-integration. *arXiv* **2023**, arXiv:2309.08142.
32. Wang, Z.; Peng, Z.; Guan, Y.; Wu, L. Manifold regularization graph structure auto-encoder to detect loop closure for visual SLAM. *IEEE Access* **2019**, *7*, 59524–59538. [CrossRef]

33. Bailey, T.; Durrant-Whyte, H. Simultaneous localization and mapping (SLAM): Part II. *IEEE Robot. Autom. Mag.* **2006**, *13*, 108–117. [CrossRef]
34. Kazerouni, I.A.; Fitzgerald, L.; Dooly, G.; Toal, D. A survey of state-of-the-art on visual SLAM. *Expert Syst. Appl.* **2022**, *205*, 117734. [CrossRef]
35. Cai, D.; Li, R.; Hu, Z.; Lu, J.; Li, S.; Zhao, Y. A comprehensive overview of core modules in visual SLAM framework. *Neurocomputing* **2024**, *590*, 127760. [CrossRef]
36. Usenko, V.; Engel, J.; Stückler, J.; Cremers, D. Direct visual-inertial odometry with stereo cameras. In Proceedings of the 2016 IEEE International Conference on Robotics and Automation (ICRA), Stockholm, Sweden, 16–21 May 2016; pp. 1885–1892.
37. Leutenegger, S.; Lynen, S.; Bosse, M.; Siegwart, R.; Furgale, P. Keyframe-based visual–inertial odometry using nonlinear optimization. *Int. J. Robot. Res.* **2015**, *34*, 314–334. [CrossRef]
38. Bloesch, M.; Omari, S.; Hutter, M.; Siegwart, R. Robust visual inertial odometry using a direct EKF-based approach. In Proceedings of the 2015 IEEE/RSJ International Conference on Intelligent Robots and Systems (IROS), Hamburg, Germany, 28 September–2 October 2015; pp. 298–304.
39. Qin, T.; Li, P.; Shen, S. Vins-mono: A robust and versatile monocular visual-inertial state estimator. *IEEE Trans. Robot.* **2018**, *34*, 1004–1020. [CrossRef]
40. Burri, M.; Nikolic, J.; Gohl, P.; Schneider, T.; Rehder, J.; Omari, S.; Achtelik, M.W.; Siegwart, R. The EuRoC micro aerial vehicle datasets. *Int. J. Robot. Res.* **2016**, *35*, 1157–1163.
41. Cho, B.S.; Moon, W.S.; Seo, W.J.; Baek, K.R. A dead reckoning localization system for mobile robots using inertial sensors and wheel revolution encoding. *J. Mech. Sci. Technol.* **2011**, *25*, 2907–2917. [CrossRef]
42. Zhu, Z.; Kaizu, Y.; Furuhashi, K.; Imou, K. Visual-inertial RGB-D SLAM with encoders for a differential wheeled robot. *IEEE Sens. J.* **2021**, *22*, 5360–5371. [CrossRef]
43. Sturm, J.; Engelhard, N.; Endres, F.; Burgard, W.; Cremers, D. A benchmark for the evaluation of RGB-D SLAM systems. In Proceedings of the 2012 IEEE/RSJ International Conference on Intelligent Robots and Systems, Vilamoura-Algarve, Portugal, 7–12 October 2012; pp. 573–580.

Disclaimer/Publisher's Note: The statements, opinions and data contained in all publications are solely those of the individual author(s) and contributor(s) and not of MDPI and/or the editor(s). MDPI and/or the editor(s) disclaim responsibility for any injury to people or property resulting from any ideas, methods, instructions or products referred to in the content.

Article

Synchronous End-to-End Vehicle Pedestrian Detection Algorithm Based on Improved YOLOv8 in Complex Scenarios

Shi Lei [1,2], He Yi [1,2] and Jeffrey S. Sarmiento [1,*]

[1] Computer Engineering Department, Batangas State University, Batangas City 4200, Philippines; 21-01463@g.batstate-u.edu.ph (S.L.); 21-08297@g.batstate-u.edu.ph (H.Y.)
[2] College of Electrical and Control Engineering, Henan University of Urban Construction, Pingdingshan City 467000, China
* Correspondence: jeffrey.sarmiento@g.batstate-u.edu.ph; Tel.: +63-908-777-7995

Citation: Lei, S.; Yi, H.; Sarmiento, J.S. Synchronous End-to-End Vehicle Pedestrian Detection Algorithm Based on Improved YOLOv8 in Complex Scenarios. *Sensors* **2024**, *24*, 6116. https://doi.org/10.3390/s24186116

Academic Editors: Enrico Meli and Bogdan Smolka

Received: 31 July 2024
Revised: 14 September 2024
Accepted: 19 September 2024
Published: 22 September 2024

Copyright: © 2024 by the authors. Licensee MDPI, Basel, Switzerland. This article is an open access article distributed under the terms and conditions of the Creative Commons Attribution (CC BY) license (https://creativecommons.org/licenses/by/4.0/).

Abstract: In modern urban traffic, vehicles and pedestrians are fundamental elements in the study of traffic dynamics. Vehicle and pedestrian detection have significant practical value in fields like autonomous driving, traffic management, and public security. However, traditional detection methods struggle in complex environments due to challenges such as varying scales, target occlusion, and high computational costs, leading to lower detection accuracy and slower performance. To address these challenges, this paper proposes an improved vehicle and pedestrian detection algorithm based on YOLOv8, with the aim of enhancing detection in complex traffic scenes. The motivation behind our design is twofold: first, to address the limitations of traditional methods in handling targets of different scales and severe occlusions, and second, to improve the efficiency and accuracy of real-time detection. The new generation of dense pedestrian detection technology requires higher accuracy, less computing overhead, faster detection speed, and more convenient deployment. Based on the above background, this paper proposes a synchronous end-to-end vehicle pedestrian detection algorithm based on improved YOLOv8, aiming to solve the detection problem in complex scenes. First of all, we have improved YOLOv8 by designing a deformable convolutional improved backbone network and attention mechanism, optimized the network structure, and improved the detection accuracy and speed. Secondly, we introduced an end-to-end target search algorithm to make the algorithm more stable and accurate in vehicle and pedestrian detection. The experimental results show that, using the algorithm designed in this paper, our model achieves an 11.76% increase in precision and a 6.27% boost in mAP. In addition, the model maintains a real-time detection speed of 41.46 FPS, ensuring robust performance even in complex scenarios. These optimizations significantly enhance both the efficiency and robustness of vehicle and pedestrian detection, particularly in crowded urban environments. We further apply our improved YOLOv8 model for real-time detection in intelligent transportation systems and achieve exceptional performance with a mAP of 95.23%, outperforming state-of-the-art models like YOLOv5, YOLOv7, and Faster R-CNN.

Keywords: vehicle pedestrian detection; complex scenes; improved YOLOv8; synchronize end-to-end

1. Introduction

With the continuous advancements in urbanization and the improvement of living standards, private car ownership has surged, alongside the growth of public transportation and shared vehicles [1]. This increase in vehicles has brought convenience, but also significant social challenges, particularly traffic congestion during rush hours and frequent traffic accidents [2]. In 2020, China reported 156,901 traffic accidents, resulting in 5225 fatalities and 195,374 injuries.

The rapid development of the Internet and artificial intelligence has made intelligent video surveillance technology feasible. Traditional single-camera monitoring systems are limited in scope, necessitating the use of multiple cameras to cover large areas such as urban centers, airports, railway stations, and residential zones. However, efficiently analyzing the vast amounts of data generated by these systems to identify specific targets and events remains a significant challenge. See Figure 1.

Figure 1. Traffic congestion and accident scene.

Artificial intelligence, particularly computer vision, offers a solution by enabling real-time and accurate vehicle and pedestrian detection [3]. This technology can help traffic management departments better understand urban traffic conditions, plan routes, and implement traffic control measures, thereby improving road use efficiency and safety [4–7]. Additionally, it enhances the effectiveness of video surveillance by providing quick and accurate information about vehicles and pedestrians, laying the groundwork for further advancements in vehicle tracking, license plate recognition, and traffic statistics [8]. This not only improves traffic safety but also presents economic opportunities.

Traditional vehicle and pedestrian detection methods include template matching, edge detection, background modeling, and variable part models. Template matching uses image processing to compare input images with template images, effectively detecting vehicles and pedestrians, even in complex environments [9,10]. Edge detection algorithms identify edge features and motion trajectories, utilizing operators like Haar, HOG, LBP, and SIFT. The Deformable Part Model (DPM) extends the HOG model by decomposing the target into sub-model components for detection and location calibration [11,12].

However, these traditional methods face challenges such as poor robustness, limited expressiveness, and high dependence on manual design, leading to detection errors in complex traffic scenes. To address these issues, this study proposes an improved algorithm based on the YOLOv8 model, enhancing multi-scale pedestrian features, strengthening detector positioning, and improving detection accuracy [13].

2. Materials and Methods

The existing methods can effectively enhance the detection effect of small-scale targets through feature fusion or reconstruction operations, but their detection performances are limited in the face of crowded scenes containing a lot of noise. In order to effectively solve the above problems, this paper focuses on enhancing their multi-scale blocking pedestrian features, strengthening the positioning ability of their detectors, and improving their detection accuracy [14]. An improved algorithm is proposed based on the YOLOv8 model. The network structure of the improved YOLOv8 algorithm proposed in this paper is shown in Figure 2.

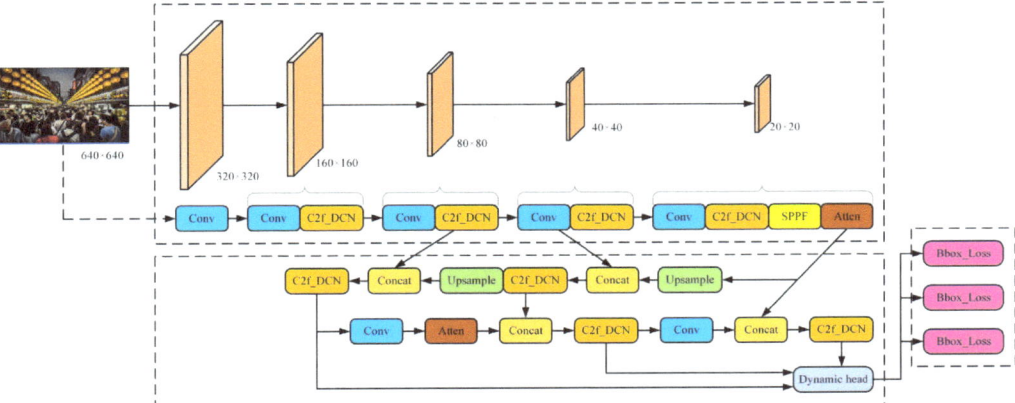

Figure 2. Network structure of improved YOLOv8 algorithm.

2.1. Improved Backbone Network

The traditional convolution operation is characterized by a fixed convolutional kernel and a limited receptive field. These characteristics allow it to capture only local features within a specific region of the input image. Consequently, this limitation hinders the ability to capture global contexts or larger patterns, which are crucial for understanding complex scenes. In scenarios involving multi-scale occlusions, the fixed nature of the convolutional kernel often leads to significant information loss, as the network struggles to detect features that vary in scale or are partially obscured [15].

To address these limitations, we propose a new method based on adaptive convolutional operations, where the convolutional kernel dynamically adjusts based on the input. Additionally, we incorporate an attention mechanism to enhance the model's ability to focus on relevant features, allowing for the more robust detection of occluded pedestrians.

2.1.1. DarkNet-53 Network

Based on YOLOv3's DarkNet-53, it adopts the traditional backbone network architecture ResNet, and adds residuals on this basis, and improves the overall performance of the model through iteration. The DarkNet-53 stacks 53 convolution modules, delivering performance comparable to the ResNet-152 network with twice the FPS. In the core network of YOLOv8, the DarkNet-53 network and the C2f module are used to achieve higher performance and weight [16]. However, the C2f model adopts the traditional convolution structure, which leads to limited perceptual domain, and cannot effectively deal with multi-scale occlusion in large-scale complex backgrounds.

2.1.2. Deformable Convolution

The deformable convolution function constructed by Dai et al. is characterized by its irregular lattice structure rather than a specific geometric figure. Compared to the conventional convolution method, the deformable convolution method increases a displacement value relative to the traditional convolution method, thus enlarging the receiving region [17–19]. The eigenmatrix is obtained by the variable convolution method

$$y(p_0) = \sum_{k=1}^{K} w_k \cdot x(p_0 + p_k + \Delta p_k) \tag{1}$$

where x and y are input and output feature maps, respectively; K and k are the total number of sampling points and enumerated sampling points, respectively. p_0 is the current pixel; w_k and p_k are the projection weight of the Kth sampling point and the Kth position of the predefined convolutional mesh sampling point, respectively. Δp_k is the offset corresponding to the Kth sampling position.

This offset Δp_k allows the convolutional kernel to adaptively sample the input feature map, effectively enlarging the receptive field and enabling the model to capture more complex spatial patterns. The algorithm can be closer to the shape and size of the human body in the actual scene, which enhances the robustness of the algorithm. See Figure 3.

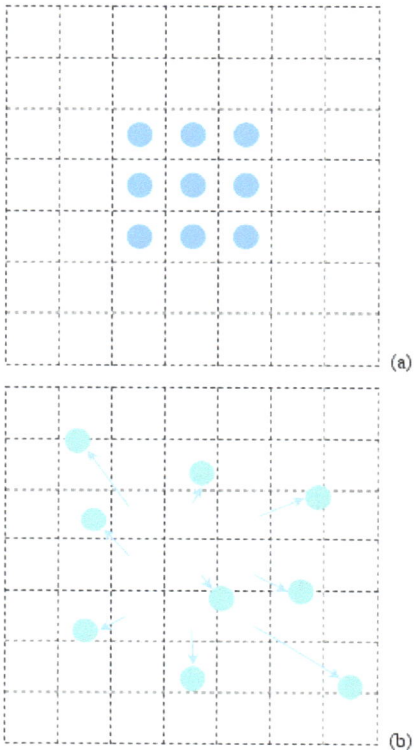

Figure 3. Sampling process ((**a**) Standard convolution sampling points; (**b**) deformable convolution sampling points).

2.1.3. C2f_DCN Module

In order to better detect pedestrians, a C2f_DCN model based on variable convolution is proposed to enhance the recognition ability of pedestrians on multiple scales [20]. Among them, the C2f_DCN model adopts a 1×1 convolution transform input characteristic, and uses a Split operation, instead of a 1×1 convolution, to realize image segmentation. By stacking multiple deformable convolutions, the perception field is improved, the jump connectivity is increased, and more abundant gradient data are obtained under the premise of reducing the number of parameters. This method has strong multi-scale characteristics.

One of the key advantages of the C2f_DCN model is its use of stacked deformable convolutions. By stacking multiple layers of deformable convolutions, the model significantly expands its receptive field, allowing it to capture more contextual information. This is particularly beneficial in detecting small or partially occluded pedestrians, where traditional models might struggle due to limited spatial awareness. Moreover, the model enhances gradient flow and connectivity by incorporating skip connections, which help preserve important feature details while reducing the risk of vanishing gradients.

Additionally, by reducing the number of parameters, the model not only becomes more computationally efficient, but also retains strong multi-scale characteristics, making it robust in varying real-world conditions. This combination of techniques enables the

C2f_DCN model to more accurately detect pedestrians of different sizes and in different states of occlusion, thus addressing key challenges in pedestrian detection. See Figure 4.

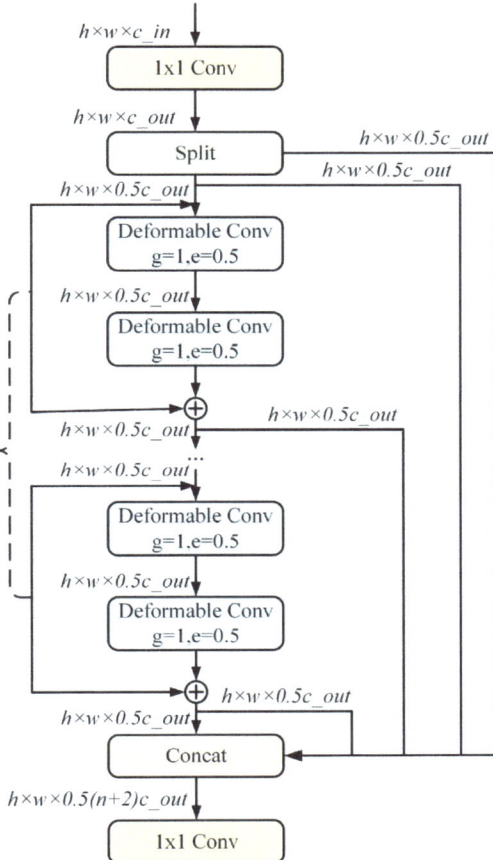

Figure 4. C2f_DCN module structure.

2.1.4. Occlusion Perceptual Attention Mechanism

The attention mechanism can quickly search the entire scene to identify the affected object, which has advantages in detecting the affected pedestrian's characteristics [21]. However, due to the adoption of the attention mechanism, it requires a large amount of calculation and computation. WOO et al. proposed a CBAM model based on spatiotemporal dual concerns. Although it has good spatiotemporal perception ability, it is computation-intensive and not lightweight [22]. Wang et al. proposed an efficient channel attribute (ECA) method, which eliminates the dimensional reduction of channel data and uses one-dimensional convolution to capture the interactions between multiple channels, reducing the complexity of the modeling process, but requiring more channel information. In the case of high density, the effect is not large [23].

The algorithm is weighted to each channel to enhance the key pedestrian characteristics and reduce the background components in the image. This paper takes "occlusion" as the research object. By comparing SE, CBAM, GAM (global attention mechanism), and Biformer, in the same network environment, the accuracy and computational efficiency of occlusion attention are better taken into account [24]. This paper proposed a visual attention mechanism based on visual coverage, through the global mean and global maximum mean of the target image, to achieve effective processing of the target image, so as to achieve effective tracking of the target. The two pool superposition operations combine the characteristics of multiple scales, improve the cognition and expression of the characteristics of small-scale pedestrians who are blocked, and reduce the information loss caused by a single pool operation. The algorithm first takes the image as a one-dimensional convolution, and then multiplies it with the matrix of the original image by the activation function of Sigmoid to obtain the final image.

The proposed visual attention mechanism, based on visual coverage, further improves the handling of occluded objects. By applying global average pooling and global maximum pooling to the target image, the mechanism achieves effective multi-scale feature extraction. The pooling operations help in better capturing the characteristics of small-scale pedestrians who are occluded, reducing the information loss that typically occurs with a single pooling operation. The algorithm processes the image using one-dimensional convolution and then multiplies it with the matrix of the original image using the Sigmoid activation function, resulting in a refined output image.

In contrast to other attention mechanisms, the occlusion perceptual attention mechanism was chosen due to its ability to prioritize occluded areas, enhancing the detection of pedestrians in cluttered and complex environments. While methods like SE and CBAM offer strong spatial and channel attention, their computational costs and limitations in high-density scenes make them less effective for our specific goal of occlusion-based pedestrian detection. The proposed mechanism improves both the robustness and efficiency in occluded pedestrian detection scenarios, which are the primary focus of this research. Occlusion perceptual attention is

$$\text{Atten}(F) = \sigma(\varphi(\text{AvgPool}(F) + \text{MaxPool}(F))) \cdot F \qquad (2)$$

where F is the input characteristic. Where $\varphi(\cdot)$ sigma is a one-dimensional convolution function. Sigma is a Sigmoid activation function. See Figure 5.

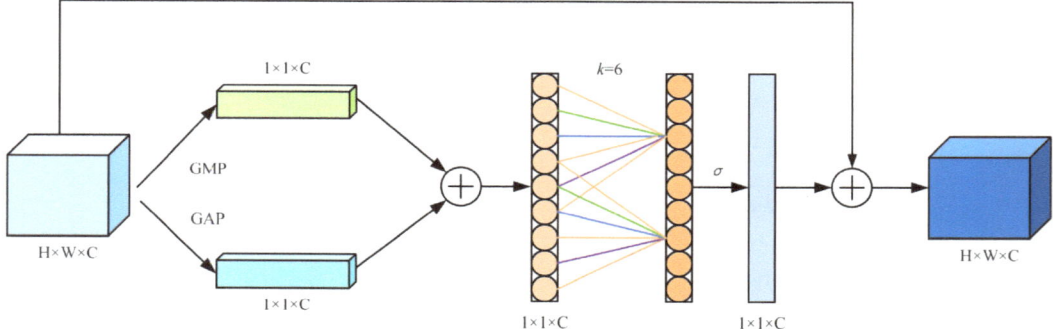

Figure 5. Occlusion perceptual attention mechanism.

2.2. Improved Dynamic Decoupling Head

The backbone network is used for image feature extraction, and the detection head is used for image recognition and location. The high-efficiency probe can effectively improve the detection effect, especially for the case of large density, and can more accurately locate the object.

2.2.1. Dynamic Detection Header

Dai et al. combined the attention mechanism with the object detection probe, and regarded the result of the backbone network as a 3-dimensional (rank × space) tensor composed of the three dimensions of "rank", "space", and "output". The proportion perception at the element level, the spatial perception attention at the spatial orientation, and the task perception attention at the output channel constitute a dynamic head structure [25]. Among them, the three dimensions of AL, AS, and AC represent the three dimensions of attention. On this basis, this paper proposes a multi-level visual information extraction method based on visual information, and uses visual information fusion technology to process visual information, so as to achieve visual information extraction based on visual information [26]. Then the features of each dimension are fused to obtain a good target detection head with a comprehensive effect. See Figure 6.

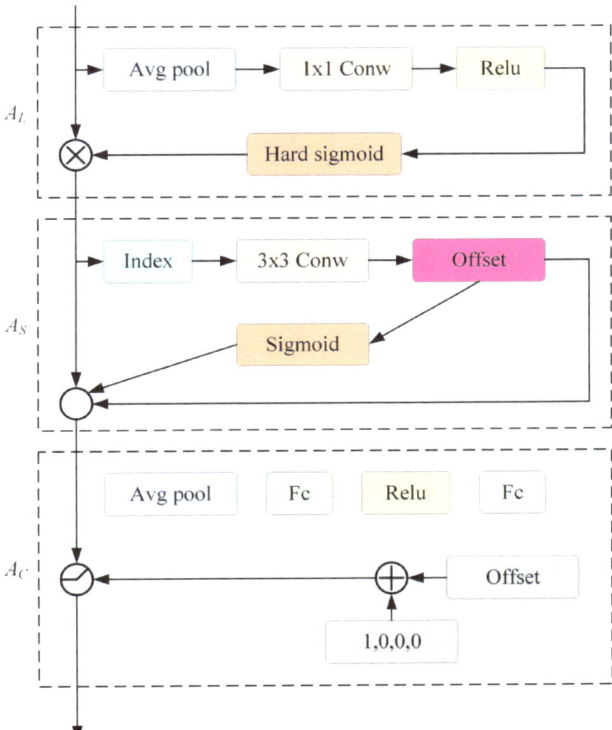

Figure 6. Structure of dynamic detection head.

2.2.2. Dynamic Decoupling Header Module

Using the moving probe for reference, the YOLOv8 probe is integrated with the characteristics of single target recognition, and a target recognition method based on dynamic decoupling is proposed. Since AS is a variable convolution, we will use deformation convolution to improve the spatial cognitive performance of the backbone, and improve the detection performance of the probe head through the scale and task-aware attention mechanism, so as to reduce the complexity of the model [27]. Since the YOLOv8 detector has only decoupled classes and regression branches, it is still necessary to decouple the characteristics of the power head. The power separation head assembly is shown in Figure 7. For a given feature tensor $T \in R L \times S \times C$, its attention can be described in the dynamic decoupling header as

$$W(T) = A_C(A_L(T) \cdot T) \cdot T \tag{3}$$

$A_C(\cdot)$ and $A_L(\cdot)$ are task-aware attention and scale-aware attention, respectively. $A_L(\cdot)$ is defined as

$$A_L(T) \cdot T = \sigma \left(f \left(\frac{1}{SC} \sum_{S,C} T \right) \right) \cdot T \tag{4}$$

$$\sigma(x) = \max\left(0, \min\left(1, \frac{x+1}{2}\right)\right) \quad (5)$$

where σ is the Hard-Sigmoid function; f (·) is a linear transformation of the 1 × 1 convolution. AC(·) is defined as

$$A_C(T) \cdot T = \max\left(\alpha^1(T) \cdot T_c + \beta^1(T), \alpha^2(T) \cdot T_c + \beta^2(T)\right) \quad (6)$$

Among them, the slice of Tc for the c channel characteristics, as well as the activation of the threshold function, is used to study control $\theta(\cdot) = \left[\alpha^1, \alpha^2, \beta^1, \beta^2\right]^T$.

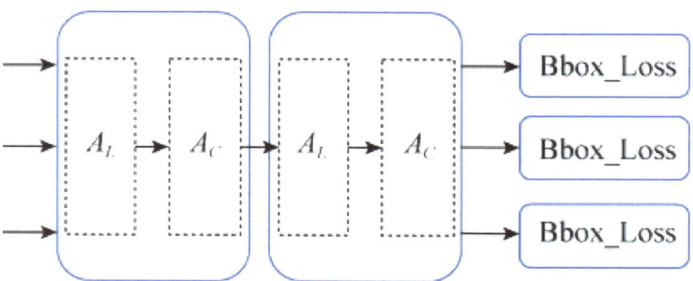

Figure 7. Dynamic decoupling header module.

2.3. Optimization of Loss Function

Because of the single type of target in complex scenes, it is necessary to further strengthen the target positioning in the process of target identification and positioning. The algorithm takes the spacing between the edge lines of the four edge boxes as the constraint function, and based on this, the parameters are estimated [28]. In general, the actual target is where the marker is, so the DFL will quickly find the region where the marker is located by increasing the two points y_i and $y_i + 1$ ($y_i \leq y \leq y_i + 1$) closest to the marker y. Therefore, the definition of DFL is

$$DFL(S_i, S_{i+1}) = -\left((y_{i+1} - y)\log(S_i) + (y - y_i)\log(S_{i+1})\right) \quad (7)$$

When counting the lost quantity, DFL only maps the coordinates of the marked frame to the marked coordinate system, but cannot determine the association between the marked coordinate system and the marked coordinate system. To solve this problem, the Wise-IoU loss method is adopted in this paper to measure the overlap loss of tag frame and anchor frame, and DFL is added to the final edge frame model.

Wise-IoU overcomes the bias in the existing IoU evaluation methods by combining the regions of the prediction frame and the actual frame, and establishes an attention-based edge box loss model [29,30]. On this basis, a new method based on edge frame is proposed, that is, the focus factor is solved, and the focusing mechanism is added on this basis, so as to achieve the purpose of focusing.

Wise-IoU loss is defined as

$$L_{WIoU} = 1 - \frac{\sum_{i=1}^{n} w_i IoU(b_i, g_i)}{\sum_{i=1}^{n} w_i} \quad (8)$$

where n is the number of object frames; b_i is the coordinate of the i th object frame; g_i is the coordinates of the actual mark box for the i th object; wi is the weight value. IoU(b_i, g_i) is the IoU value between the I-th object box and the real marked box.

The final regression loss obtained after combining the above two losses is

$$L_{reg} = \lambda \cdot DFL + \mu \cdot L_{WIoU} \qquad (9)$$

The determination of the weight coefficients λ and μ is based on the regression loss weight setting of YOLOv8. A high IoU value is very important for accurate target location and detection, so a large weight is needed. DFL can easily cause overfitting problems in model training, which affects the generalization ability of the model, so smaller weights are needed. Therefore, through the relevant experimental analysis, $\lambda = 1/6$, $\mu = 5/6$. The combined regression loss improves the training efficiency of the model.

3. End-to-End Target Search Algorithm

The end-to-end target search algorithm refers to the end-to-end network training solution of target detection in the same model in the way of multi-task head.

One of the most representative is OIM. The structure of the model is shown in Figure 8. On this basis, a method of target detection based on the fast convolutional neural network is proposed. For panoramic images, the framework convolutional neural network is used in the OIM network to extract the feature map, the existing pedestrian candidate network is used to construct the edge frame, and the ROI pool layer is used to obtain the individual feature range. Finally, this data is used as the final feature extractor. The extracted feature information is shared through the pedestrian detection network and pedestrian re-marking network. First, it is mapped to L2 normalized 256-dimensional subspace, and then it is learned using sample matching costs.

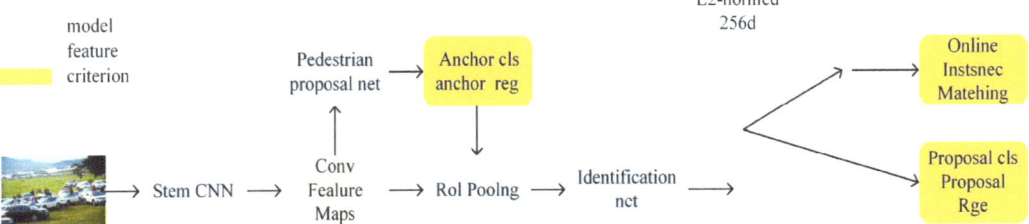

Figure 8. Schematic diagram of OIM structure.

The OIM algorithm laid the foundation for the end-to-end target search algorithm, and the subsequent end-to-end target search network model basically followed the network architecture of OIM. The biggest contribution of OIM is realizing the end-to-end training of the target search model, saving the training time and reducing the training difficulty of the model. However, the OIM network does not propose a solution to the problem of goal contradiction between the two subtasks in the target search task. In addition, it is a fine-grained task that requires individual division of similar people, and the two subtasks of the OIM network are parallel structures. The pedestrian features of the input network come from the low-quality bounding box generated by the RPN of the candidate region generation network with a lot of background information, which has a great influence on the result.

To solve this problem, this paper intends to adopt the embedded decomposition algorithm based on the local least squares support vector to transform the pedestrian information into vectors, so as to realize the effective recognition of complex scenes. The research results of this paper can effectively improve the retrieval efficiency and improve the retrieval precision. On this basis, this paper intends to extend the above algorithms from the region level to the pixel level, reduce the influence of non-matching factors on model identification performance, and improve the efficiency of multi-task learning. Figure 9 shows the overall network structure.

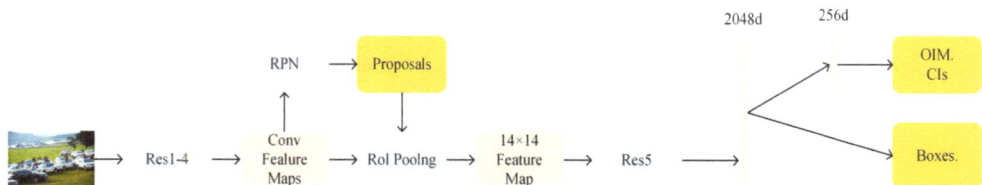

Figure 9. Schematic diagram of NAE structure.

In order to obtain higher quality candidate boxes, Li et al. adjusted NAE as the baseline network structure and proposed the SeqNet algorithm [31]. The main idea is to leverage the Faster R-CNN as a stronger RPN to provide fewer but more accurate and higher quality candidate boxes that provide more differentiated pedestrian feature embeddings. As shown in Figure 10, SeqNet consists of two serialized header networks that solve pedestrian detection and pedestrian re-recognition problems, respectively. The first standard Faster R-CNN headers were used to generate accurate candidate boxes. A second, unmodified baseline header is used to further fine-tune these candidate boxes and extract their identification characteristics.

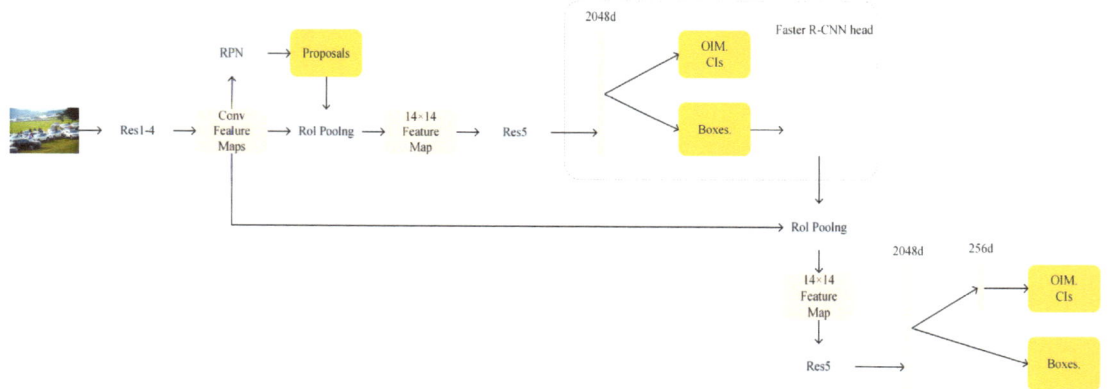

Figure 10. SeqNet structure diagram.

The network structure of the above end-to-end target search model is built on the basis of Faster R-CNN. The pedestrian candidate frame generated by the RPN network is used to extract pedestrian features one by one from the feature maps in the trunk network. This network based on candidate frame can better locate pedestrians and extract high-quality pedestrian features. But, the dense candidate frame also leads to a lot of candidate frame coordinate calculation problems.

Therefore, Yan et al. proposed the first Feature-Aligned Person Search Network (AlignPS) without an anchor frame [32–34]. The AlignPS network directly abandons the RPN network. The baseline network is constructed by taking FCOS, a frameless network in the target field, and its structure is shown in Figure 11. Since there is no RPN network, there is no problem with calculating a large number of candidate boxes, but due to the absence of operations such as ROI-Align, the pedestrian feature embedment of AlignPS, an unanchored box model, must be learned directly from the feature map without explicit regional alignment. At the same time, in order to learn multi-scale features in order to adapt to the scale changes of the target, the AlignPS model uses the improved AFA of the feature pyramid FPN as a feature fusion module. In order to maintain the consistency of the same pedestrian features at different scales, the AFA network only outputs the feature map of the last layer as an input for subsequent tasks. In response to the misalignment of tasks, AlignPS converts the original detection priority to the re-identification priority and proposes the principle of the "Re-id priority", which means that the output of the backbone network first imposes a pedestrian re-identification loss for supervisory training, and then the target feature is embedded into the input detector to complete the target detection task.

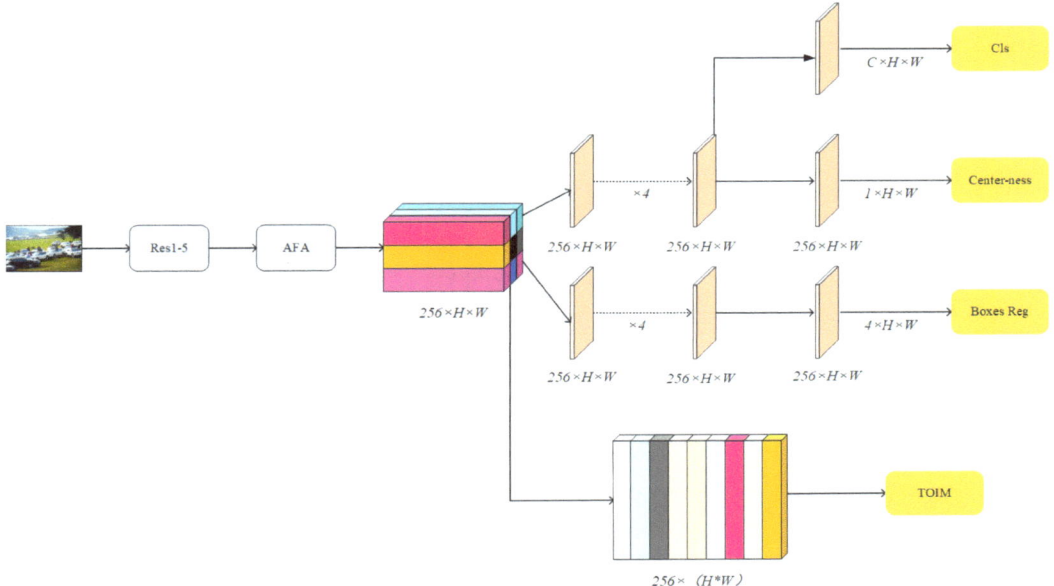

Figure 11. AlignPS structure diagram.

4. Attention Mechanism

The attention mechanism can perform weighted processing when processing input data strengthen important information attention and ignore unnecessary information when executing deep convolutional neural network tasks. Improved YOLOv8 adds ECA attention mechanisms to the VGG module of the backbone network to improve the accuracy and performance of the model by adjusting attention to input data.

4.1. Attention Mechanism of ECA Channel

Channel attention mechanisms have great potential to improve the performance of neural networks, but most of them are designed into complex structures to achieve better performance, which undoubtedly increases the complexity of the model. ECA attention module has a simple structure but can bring obvious gain effect. By analyzing the attention module of the SE channel, it is found that it is inefficient to obtain the correlation between all channels at the same time. Therefore, the ECA model is a further improvement of the SE model. Among them, the SE module uses a Fully Connected (FC) layer to represent channel interest, the ECA uses 1×1 convolution to represent channel interest, and uses one-dimensional convolution to realize information interaction among multiple channels, reducing the parameter values of each channel in the model. In each convolution process, only some channels play a certain role. In this way, the information exchange between different channels can be better realized. Based on the different channel scales, the size of the convolutional core can be flexibly adjusted to realize the mutual influence among the layers with a large number of channels, thus greatly improving the channel learning speed.

The ECA attention module is shown in Figure 12. Firstly, the spatial characteristics of the image are extracted, and the $1 \times 1 \times C$ image is obtained by using the whole average pooling method. On this basis, the 1D convolution model is used to solve the dimension of the 1D convolution function, and the channel interest of $1 \times 1 \times C$ is obtained, and the original input feature map is fused to obtain the image with the channel interest. The adaptive calculation formula for the convolution core size is

$$k = \Phi(C) = \left| \frac{\log_2 C + b}{\gamma} \right|_{odd} \tag{10}$$

where k represents the size of the convolution kernel; C indicates the number of channels. γ and b are the parameters of the adaptive calculation adjustment. $|\cdot|_{odd}$: The value of k is odd

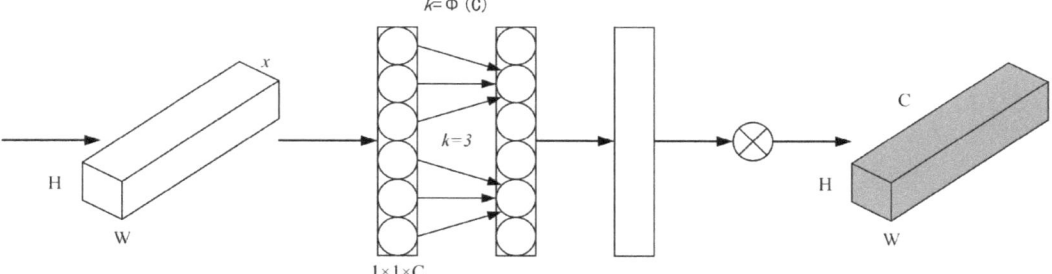

Figure 12. ECA attention module.

4.2. eSE Attention Mechanism

The SE focus module dynamically adjusts the weights of each channel, thereby improving the overall performance of the system. In order to avoid the high modeling complexity, the first order FC uses the reduced parameter r to reduce the input characteristic channel C to C/r. Second order FC will reduce the channel length to the initial channel size C. During this period, the channel dimension decreases, and the channel information is lost. The eSE model further modifies the model, that is, there is only one FC level and only one C-level channel, from the original two FC levels to C level. Therefore, the robustness of the algorithm is improved on the premise that the channel information is not lost.

In the C2f module of YOLOv8, the input is first separated through the first convolution layer and is divided into two parts: one part directly passes through n Bottleneck, and the other part is separated again after each operation layer to create a jump connection. The multiple feature maps of all the branches are collected in the eSE module. As shown in Figure 13, after the eSE module is globally average pooled and fully connected, the feature map learns and outputs the channel attention information, and fuses it into the feature map in the form of elements through feature mapping.

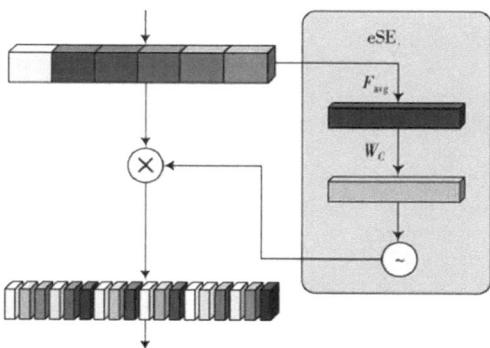

Figure 13. eSE module workflow.

5. Results

5.1. Experimental Data Set

The dataset used for evaluation is the widely adopted Cityscapes dataset, consisting of 5000 finely annotated urban street images. It captures complex urban traffic scenes, including varied lighting conditions, occlusions, and dense object distributions, making it suitable for testing vehicle and pedestrian detection algorithms. The dataset was divided into 2975 training samples, 500 validation samples, and 1525 testing samples. The diversity of the dataset enables comprehensive evaluation of our proposed algorithm in real-world conditions.

5.2. Experimental Environment and Parameter Configuration

Test environment: Ubuntu1804 OS, two NVIDIA TITAN Xp, Python3.8 training platform, PyTorch deep learning framework, YOLOv8s pattern library. Set the training parameters for this pattern to the Epoch size of 200 and the batch size of 32, and all other parameters are at default.

5.3. Evaluation Index

This paper mainly uses the following evaluation indexes to evaluate the results of the algorithm.

The calculation formulas for accuracy (P) and recall (R) are shown in Equations (11) and (12):

$$P = \frac{T_{TP}}{T_{TP} + F_{TP}} \tag{11}$$

$$R = \frac{T_{TP}}{T_{TP} + F_{FN}} \tag{12}$$

The calculation formula of the average accuracy mAP is shown in Equation (13):

$$m = \frac{1}{N} \sum_{n=0}^{N} \int_{0}^{1} P_n(r) dr \tag{13}$$

where N represents the number of categories of detection targets; P_n indicates the AP value of a certain category.

mAP@0.5 sets the IoU to 0.5, calculates the AP for each class of images, and finally averages all the classes, known as mAP.

Frame rate (FPS) represents the number of images a model can process per second, which is usually used to measure the real-time performance of the model, supplemented by the number of parameters and amount of computation (GFLOPs).

5.4. Experimental Results

Our proposed algorithm showed significant improvements in both vehicle and pedestrian detection compared to state-of-the-art models. The experimental results are shown in Table 1 below.

Table 1. Comparison of algorithm test results.

Algorithm	Vehicle R/%	Pedestrian R/%	Vehicle P/%	Pedestrian P/%	Map/%
Faster R-CNN	92.52	88.13	88.47	93.04	90.89
YOLOv5	91.93	83.77	86.94	92.78	88.86
YOLOv7	93.44	84.28	87.06	93.33	89.23
SORT	91.57	82.47	85.13	92.32	87.08
YOLOv8	94.31	85.21	85.92	94.89	89.61
Our algorithm	97.19	88.65	96.02	98.84	95.23

Our experimental results demonstrate that the proposed algorithm outperforms both the baseline YOLOv8 and several state-of-the-art detection models, including YOLOv5, YOLOv7, and Faster R-CNN. Our algorithm achieves notable improvements in detection accuracy for both vehicles and pedestrians. Specifically, our method achieves a vehicle recall of 97.19%, surpassing YOLOv8's 94.31% and Faster R-CNN's 92.52%. Additionally, pedestrian precision reaches 96.02%, a significant enhancement over YOLOv8's 94.89% and Faster R-CNN's 93.04%. These improvements can be attributed to the enhanced feature extraction and attention mechanisms that allow the model to more effectively capture complex scene dynamics. Moreover, the proposed algorithm achieves a mean average precision (mAP) of 95.23%, outperforming YOLOv8 (89.61%) and Faster R-CNN (90.89%). The enhanced performance of our model, especially in crowded scenes, underscores its robustness and applicability in real-world traffic environments.

5.5. Optimization of Loss Function

The best performances after the simplification and optimization of the original network model of YOLOv8 and the network model of this paper were respectively trained and tested, and comparative experiments were conducted. The experimental results are shown in Table 2 below. See Figure 14.

Table 2. Experimental results before and after loss function optimization.

Algorithm	Loss Function	Training Duration (Days)	Common Test Set AP (%)	Intensive Test Set AP (%)	Detection Speed Fps
Faster R-CNN	before	4	92.81	71	15
	after	4.5	93.73	82	15.1
YOLOv5	before	3	91.52	68	40
	after	3.5	92.54	75	40.1
YOLOv7	before	3	92.11	69.5	45
	after	3.5	93.09	79	45.2
SORT	before	N/A	88.44	67.82	47.32
	after	N/A	89.57	77.67	47.39
YOLOv8	before	3	92.74	72	38.67
	after	3.5	93.86	84.2	38.71
Our algorithm	before	2	94.86	71.44	41.52
	after	2.5	96.02	90.3	41.46

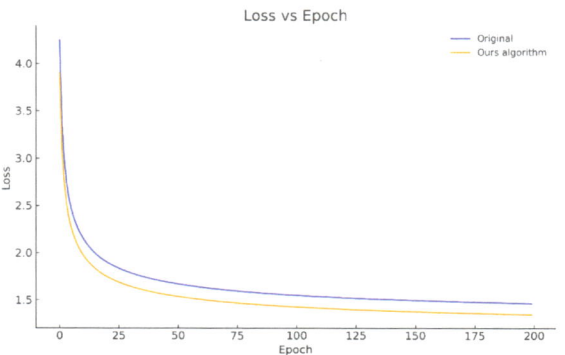

Figure 14. Comparison of training loss curves.

The table illustrates the performance improvements of various algorithms before and after the application of optimized loss functions. Our algorithm demonstrates the most significant gains, with the common test set AP increasing from 94.86% to 96.02%, and the intensive test set AP showing a dramatic rise from 71.44% to 90.3%. Despite the substantial improvements in accuracy, our model maintains a real-time detection speed, with the FPS slightly adjusting from 41.52 to 41.46. Compared to other algorithms such as YOLOv8 and SORT, our algorithm achieves a better balance between detection accuracy and speed.

However, despite the improvements, our algorithm faces challenges under conditions with extreme lighting variations or highly complex occlusions. Further research is needed to enhance robustness under these conditions and to reduce computational overhead for deployment on edge devices. Future work will focus on refining the attention mechanisms and expanding the model's capability to handle a wider range of environmental factors.

6. Conclusions

The experiment verifies that the pedestrian and vehicle detection used in this paper is feasible, which not only improves the detection accuracy, but also meets the real-time requirements. The detection effect diagram of the final algorithm model in this paper is shown in Figure 15.

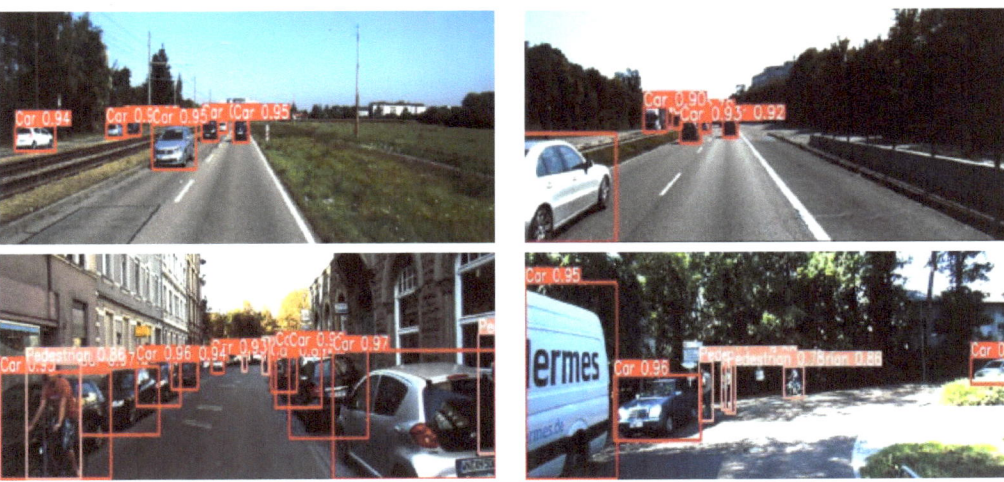

Figure 15. Model detection effect diagram.

For vehicles and pedestrians in complex environments, this paper designs a new algorithm based on improved YOLOv8. In this project, an adaptive neural network model is added to the residual matrix to enhance its characterization of features, and a depth-separable convolution model is adopted to replace the traditional 3×3 convolution to reduce the parameters of the neural network. Experiments show that, compared to YOLOv8, this method has improved accuracy and other indexes. However, despite the improvements, our algorithm faces challenges under conditions with extreme lighting variations or highly complex occlusions. Further research is needed to enhance robustness under these conditions and to reduce computational overhead for deployment on edge devices. Future work will focus on refining the attention mechanisms and expanding the model's capability to handle a wider range of environmental factors.

In conclusion, the algorithm proposed in this study provides substantial improvements in both detection accuracy and speed, making it highly applicable for intelligent transportation systems and autonomous driving technologies.

Author Contributions: Conceptualization, S.L. and J.S.S.; methodology, S.L.; software, S.L.; validation, S.L. and H.Y.; formal analysis, S.L.; data curation, S.L.; writing—original draft preparation, S.L.; writing—review and editing, S.L.; visualization, S.L.; supervision, J.S.S.; project administration, J.S.S.; All authors have read and agreed to the published version of the manuscript.

Funding: This research received no external funding.

Institutional Review Board Statement: Not applicable.

Informed Consent Statement: Not applicable.

Data Availability Statement: The original contributions presented in the study are included in the article, further inquiries can be directed to the corresponding author.

Conflicts of Interest: The authors declare no conflict of interest.

References

1. Xu, Y.; Wen, M.; He, W.; Wang, H.; Xue, Y. An improved multi-scale and knowledge distillation method for efficient pedestrian detection in dense scenes. *J. Real Time Image Process.* **2024**, *21*, 126. [CrossRef]
2. Fang, Y.; Pang, H. An Improved Pedestrian Detection Model Based on YOLOv8 for Dense Scenes. *Symmetry* **2024**, *16*, 716. [CrossRef]
3. Wang, B.; Li, Y.Y.; Xu, W.; Wang, H.; Hu, L. Vehicle–Pedestrian Detection Method Based on Improved YOLOv8. *Electronics* **2024**, *13*, 2149. [CrossRef]
4. Liang, W. Research on a vehicle and pedestrian detection algorithm based on improved attention and feature fusion. *Math. Biosci. Eng. MBE* **2024**, *21*, 5782–5802. [CrossRef]
5. Wang, L.; Bai, J.; Wang, P.; Bai, Y. Research on Pedestrian Detection Algorithm in Industrial Scene Based on Improved YOLOv7-Tiny. *IEEJ Trans. Electr. Electron. Eng.* **2024**, *19*, 1203–1215. [CrossRef]
6. Yang, J.; Huang, W. Pedestrian and vehicle detection method in infrared scene based on improved YOLOv5s model. *Autom. Mach. Learn.* **2024**, *5*, 90–96.
7. Tahir, N.U.A.; Long, Z.; Zhang, Z.; Asim, M.; ELAffendi, M. PVswin-YOLOv8s: UAV-Based Pedestrian and Vehicle Detection for Traffic Management in Smart Cities Using Improved YOLOv8. *Drones* **2024**, *8*, 84. [CrossRef]
8. Gao, L.; Yang, Y. A Method of Lightweight Pedestrian Detection in Rainy and Snowy Weather Based on Improved YOLOv5. *Acad. J. Comput. Inf. Sci.* **2024**, *7*, 1–7.
9. Han, W.; He, N.; Wang, X.; Sun, F.; Liu, S. IDPD: Improved deformable-DETR for crowd pedestrian detection. *Signal Image Video Process.* **2023**, *18*, 2243–2253. [CrossRef]
10. Jin, Y.; Lu, Z.; Wang, R.; Liang, C. Research on lightweight pedestrian detection based on improved YOLOv5. *Math. Models Eng.* **2023**, *9*, 178–187. [CrossRef]
11. Sun, J.; Wang, Z. Vehicle And Pedestrian Detection Algorithm Based on Improved YOLOv5. *IAENG Int. J. Comput. Sci.* **2023**, *50*, 1401–1409.
12. Zhang, Q.; Liu, Y.; Zhang, Y.; Zong, M.; Zhu, J. Improved YOLOv3 Integrating SENet and Optimized GIoU Loss for Occluded Pedestrian Detection. *Sensors* **2023**, *23*, 9089. [CrossRef] [PubMed]
13. Li, J.; Xu, Z.; Xu, L. Vehicle and pedestrian detection method based on improved YOLOv4-tiny. *Optoelectron. Lett.* **2023**, *19*, 623–628. [CrossRef]
14. Wei, J.; Su, S.; Zhao, Z.; Tong, X.; Hu, L.; Gao, W. Infrared pedestrian detection using improved UNet and YOLO through sharing visible light domain information. *Measurement* **2023**, *221*, 113442. [CrossRef]
15. Li, C.; Wang, Y.; Liu, X. An Improved YOLOv7 Lightweight Detection Algorithm for Obscured Pedestrians. *Sensors* **2023**, *23*, 5912. [CrossRef]
16. Gao, F.; Cai, C.; Jia, R.; Hu, X. Improved YOLOX for pedestrian detection in crowded scenes. *J. Real Time Image Process.* **2023**, *20*, 24. [CrossRef]
17. Li, M.L.; Sun, G.B.; Yu, J.X. A Pedestrian Detection Network Model Based on Improved YOLOv5. *Entropy* **2023**, *25*, 381. [CrossRef]
18. Zuo, X.; Wang, Z.; Shen, J.; Yang, W. Improving multispectral pedestrian detection with scale-aware permutation attention and adjacent feature aggregation. *IET Comput. Vis.* **2022**, *17*, 726–738. [CrossRef]
19. Hong, F.; Lu, C.H.; Tao, W.; Jiang, W. Improved SSD Model for Pedestrian Detection in Natural Scene. *Wirel. Commun. Mob. Comput.* **2022**, *2022*, 1500428. [CrossRef]
20. Zhang, Y.; Xu, L.; Zhang, Y. Research on hierarchical pedestrian detection based on SVM classifier with improved kernel function. *Meas. Control* **2022**, *55*, 1088–1096. [CrossRef]
21. Kong, H.; Chen, Z.; Yue, W.; Ni, K. Improved YOLOv4 for Pedestrian Detection and Counting in UAV Images. *Comput. Intell. Neurosci.* **2022**, *2022*, 6106853. [CrossRef] [PubMed]
22. Devi, S.; Thopalli, K.; Malarvezhi, P.; Thiagarajan, J.J. Improving Single-Stage Object Detectors for Nighttime Pedestrian Detection. *Int. J. Pattern Recognit. Artif. Intell.* **2022**, *36*, 2250034. [CrossRef]

23. Kamalasanan, V.; Feng, Y.; Sester, M. Improving 3d pedestrian detection for wearable sensor data with 2d human pose. *ISPRS Ann. Photogramm. Remote Sens. Spat. Inf. Sci.* **2022**, *V-4-2022*, 219–226. [CrossRef]
24. Panigrahi, S.; Raju, U.S.N. InceptionDepth-wiseYOLOv2: Improved implementation of YOLO framework for pedestrian detection. *Int. J. Multimed. Inf. Retr.* **2022**, *11*, 409–430. [CrossRef]
25. Shao, Y.; Zhang, X.; Chu, H.; Zhang, X.; Zhang, D.; Rao, Y. AIR-YOLOv3: Aerial Infrared Pedestrian Detection via an Improved YOLOv3 with Network Pruning. *Appl. Sci.* **2022**, *12*, 3627. [CrossRef]
26. Tan, F.; Xia, Z.; Ma, Y.; Feng, X. 3D Sensor Based Pedestrian Detection by Integrating Improved HHA Encoding and Two-Branch Feature Fusion. *Remote Sens.* **2022**, *14*, 645. [CrossRef]
27. Mao, Y. A pedestrian detection algorithm for low light and dense crowd Based on improved YOLO algorithm. *MATEC Web Conf.* **2022**, *355*, 03020. [CrossRef]
28. Assefa, A.A.; Tian, W.; Acheampong, K.N.; Aftab, M.U.; Ahmad, M. Small-Scale and Occluded Pedestrian Detection Using Multi Mapping Feature Extraction Function and Modified Soft-NMS. *Comput. Intell. Neurosci.* **2022**, *2022*, 9325803. [CrossRef]
29. Yang, S.; Chen, Z.; Ma, X.; Zong, X.; Feng, Z. Real-time high-precision pedestrian tracking: A detection–tracking–correction strategy based on improved SSD and Cascade R-CNN. *J. Real Time Image Process.* **2021**, *19*, 287–302. [CrossRef]
30. Hui, L.; Keyang, C. Pedestrian detection algorithm based on improved muti-scale feature fusion. *J. Phys. Conf. Ser.* **2021**, *2078*, 012008.
31. Hao, S.; He, T.; Ma, X.; Zhang, X.; Wu, Y.; Wang, H. KDBiDet: A Bi-Branch Collaborative Training Algorithm Based on Knowledge Distillation for Photovoltaic Hot-Spot Detection Systems. *IEEE Trans. Instrum. Meas.* **2024**, *73*, 3504615. [CrossRef]
32. Hao, S.; An, B.; Ma, X.; Sun, X.; He, T.; Sun, S. PKAMNet: A Transmission Line Insulator Parallel-Gap Fault Detection Network Based on Prior Knowledge Transfer and Attention Mechanism. *IEEE Trans. Power Deliv.* **2023**, *38*, 3387–3397. [CrossRef]
33. Hao, S.; Gao, S.; Ma, X.; An, B.; He, T. Anchor-free infrared pedestrian detection based on cross-scale feature fusion and hierarchical attention mechanism. *Infrared Phys. Technol.* **2023**, *131*, 104660. [CrossRef]
34. Cityscapes 3D Benchmark Online. Available online: https://www.cityscapes-dataset.com/Cityscapes (accessed on 18 September 2024).

Disclaimer/Publisher's Note: The statements, opinions and data contained in all publications are solely those of the individual author(s) and contributor(s) and not of MDPI and/or the editor(s). MDPI and/or the editor(s) disclaim responsibility for any injury to people or property resulting from any ideas, methods, instructions or products referred to in the content.

Article

Time Series Prediction of Gas Emission in Coal Mining Face Based on Optimized Variational Mode Decomposition and SSA-LSTM

Jingzhao Zhang, Yuxin Cui *, Zhenguo Yan, Yuxin Huang, Chenyu Zhang, Jinlong Zhang, Jiantao Guo and Fei Zhao

College of Safety Science and Engineering, Xi'an University of Science and Technology, Xi'an 710054, China; 18954603644@163.com (J.Z.); yanzg@xust.edu.cn (Z.Y.); 20120089017@stu.xust.edu.cn (Y.H.); z18234611507@163.com (C.Z.); 22220226186@stu.xust.edu.cn (J.Z.); g18339899103@163.com (J.G.); 13571142857@163.com (F.Z.)
* Correspondence: 22220089044@stu.xust.edu.cn

Abstract: The accurate prediction of gas emissions has important guiding significance for the prevention and control of gas disasters in order to further improve the prediction accuracy of gas emissions in the mining face. According to the absolute gas emission monitoring data of the 1417 working face in a coal mine in Shaanxi Province, a GA-VMD-SSA-LSTM gas emission prediction model (GVSL) based on genetic algorithm (GA)-optimized variational mode decomposition (VMD) and sparrow search algorithm (SSA)-optimized long short-term memory (LSTM) is proposed. Firstly, a VMD evaluation standard for evaluating the amount of decomposition loss is proposed. Under this standard, the GA is used to find the optimal parameters of the VMD. Then, the SSA is used to optimize the key parameters of the LSTM to establish a GVSL prediction model. The model predicts each component and finally superimposes the prediction results for each component to obtain the final gas emission result. The results show that the accuracy of the evaluation indexes of the GVSL model and VMD-LSTM model, as well as the SSA-LSTM model and Gaussian process regression (GPR) model, are compared and analyzed horizontally and vertically under three scenarios with prediction sets of 121,94 and 57 groups. The GVSL model has the best prediction effect, and its fitting degree R2 values are 0.95, 0.96, and 0.99, which confirms the effectiveness of the proposed GVSL model for the time series prediction of gas emission in the mining face.

Keywords: time series prediction of gas emissions; variational mode decomposition; sparrow search algorithm; long short-term memory

Citation: Zhang, J.; Cui, Y.; Yan, Z.; Huang, Y.; Zhang, C.; Zhang, J.; Guo, J.; Zhao, F. Time Series Prediction of Gas Emission in Coal Mining Face Based on Optimized Variational Mode Decomposition and SSA-LSTM. *Sensors* **2024**, *24*, 6454. https://doi.org/10.3390/s24196454

Academic Editor: Ram M. Narayanan

Received: 13 September 2024
Revised: 29 September 2024
Accepted: 2 October 2024
Published: 6 October 2024

Copyright: © 2024 by the authors. Licensee MDPI, Basel, Switzerland. This article is an open access article distributed under the terms and conditions of the Creative Commons Attribution (CC BY) license (https://creativecommons.org/licenses/by/4.0/).

1. Introduction

The accurate prediction of gas emissions has important guiding significance for preventing and controlling gas disasters [1]. Improving the accuracy of gas emission prediction is one of the important research branches of gas disaster prevention [2,3].

The static models established by traditional prediction methods, such as the mine statistics method, fractional source prediction method, and neural network prediction [4], fail to consider that gas emission is a dynamic nonlinear system [5]. With the deepening of the research, it cannot meet the requirement of prediction accuracy. In order to improve the prediction accuracy of the model, many scholars introduced influencing factors of gas emissions (coal seam gas content, coal thickness, layer spacing, etc.). A multi-index gas emission prediction model was established [6–9]. WANG Yanbin analyzed the factors affecting gas emissions in the working face and then predicted the gas emissions based on the PCA-PSO-ELM model. The results show that the prediction effect of the model is better than that of the random forest and extreme learning machine models. Although this method improves the prediction accuracy, it has two disadvantages: First, most mines cannot provide detailed data such as coal thickness and adjacent layer thickness. Second, most prediction models are unable to forecast for a long duration and on a large scale [10–12]. Therefore, many scholars have explored and studied the timing prediction model of gas emissions based on the data on gas emissions [13–15].

At present, the timing series prediction model of gas emissions mostly adopts the combination prediction method based on signal decomposition, such as wavelet decomposition, empirical mode decomposition, and variational mode decomposition [16–19]. Among them, variational mode decomposition has better noise robustness than other signal decomposition methods [20]. Some scholars have verified its superiority in the fields of power load and wind speed forecasting [21–23]. However, the following problem remains:

(1) The effect of variational mode decomposition mainly depends on the setting of decomposition number k and quadratic penalty factor α, but its value is often set by experience and lacks selection criteria, so it is difficult to guarantee the decomposition effect [24,25].

(2) The prediction model plays an important role in the prediction of gas emissions. Benefiting from the advantages of abstracting and extracting features from input signals layer by layer to dig out deeper potential rule information, the deep learning model has been gradually applied to the field of timing prediction [26]. As a deep learning model, the LSTM model introduces the concept of a time sequence into the network structure, which provides a good effect on time sequence prediction and has achieved good application effects in the fields of power load prediction and photovoltaic power prediction [27,28]. However, there are few studies in the field of gas emission time sequence prediction.

For the above problems, a GA-VMD-SSA-LSTM (GVSL) gas emission prediction model based on genetic algorithm (GA)-optimized variational mode decomposition (VMD) and sparrow search algorithm (SSA)-optimized long short-term memory (LSTM) is proposed.

2. Materials and Methods

2.1. Variational Mode Decomposition (VMD)

Different from the EMD, LMD, and EEMD decomposition methods of recursive mode decomposition, VMD is a new non-recursive and adaptive signal decomposition method, which was proposed in 2014 [29]. Thanks to the introduction of the variational model, it effectively avoids the endpoint and modal aliasing effects in the recursive mode decomposition method [30].

Based on the concepts of the Wiener filter, Hilbert transform, signal parsing, mixing, and heterodyne demodulation, the steps of VMD construction are proposed as follows [31]:

(1) In order to evaluate the modal bandwidth, the Hilbert transform is introduced to transform the problem into a constrained variational problem. The equation is shown in (1):

$$\min_{\{u_k\},\{\omega_k\}} \left\{ \sum_k \left\| \partial_t \left[\left(\delta(t) + \frac{j}{\pi t} \right) * u_k(t) \right] e^{-j\omega_k t} \right\|_2^2 \right\} \\ s.t. \sum_k u_k = f(t) \qquad (1)$$

In the formula, u_k is the set of mode decomposition components, ω_k is the set of center frequencies corresponding to the decomposition components, k is the number of VMD decompositions, and $f(t)$ is the input signal to be decomposed.

(2) By introducing the Lagrange multiplier and quadratic penalty factor, the constraint form is transformed into an unconstrained form. The equation is shown in (2):

$$L(\{u_k\},\{\omega_k\},\lambda) := \alpha \sum_k \left\| \partial_t \left[\left(\delta(t) + \frac{j}{\pi t} \right) * u_k(t) \right] e^{-j w_k t} \right\|_2^2 \\ + \left\| f(t) - \sum_k u_k(t) \right\|_2^2 + \left\langle \lambda(t), f(t) - \sum_k u_k(t) \right\rangle \qquad (2)$$

where L is the augmented Lagrangian, λ is the Lagrangian multiplier, and α is the quadratic penalty factor.

(3) The alternating direction method of multipliers (ADMM) is introduced to find the saddle point of the augmented Lagrangian and solve the original minimization problem. The optimal solutions of u_k and ω_k in Equation (1) are obtained, and the calculation process is as follows, as shown in Equations (3) and (4).

$$\hat{u}_k^{n+1}(\omega) = \frac{\hat{f}(\omega) - \sum_{i \neq k} \hat{u}_i(\omega) + \frac{\hat{\lambda}(\omega)}{2}}{1 + 2\alpha(\omega - \omega_k)^2} \qquad (3)$$

$$\omega_k^{n+1} = \frac{\int_0^\infty \omega \left|\widehat{u}_k(\omega)\right|^2 d\omega}{\int_0^\infty \left|\widehat{u}_k(\omega)\right|^2 d\omega} \qquad (4)$$

In the equation: $\hat{u}_k^{n+1}(\omega)$ is the Wiener filter of the current residual $\hat{f}(\omega) - \sum_{i \neq k} \hat{u}_i(\omega) + \frac{\hat{\lambda}(\omega)}{2}$, and ω_k^{n+1} is the center of gravity of the current modal power spectrum.

2.2. Variational Mode Decomposition Based on Genetic Algorithm Optimization

Among the VMD decomposition parameters, k determines the number of mode decompositions, and α affects the fidelity and effect of the modal components. In the field of time series prediction, the selection of parameter values is mainly based on an empirical setting and spectrum analysis setting. The former lacks a theoretical basis and makes it difficult to ensure decomposition quality, while the latter is limited by the absolute gas emission data collected in this paper, which cannot be analyzed for its spectrum and, thus, makes it difficult to determine the parameter values. Therefore, the GA algorithm was introduced to optimize the parameters of k and α in VMD to ensure the decomposition effect of absolute gas emission data in VMD.

In an ideal situation, the data reconstructed by the VMD decomposition component is the same as the original data, but there is often a decomposition loss in the actual decomposition. In order to evaluate this part of the loss, the root mean square error (*RMSE*) is introduced, as shown in Equation (5).

$$RMSE = \sqrt{\frac{1}{n} \sum_{i=1}^n (y - y')^2} \qquad (5)$$

In the formula, n is the number of samples collected for the absolute gas emission, y is the measured value of absolute gas emission, and y' is the reconstructed value of gas emission of the VMD decomposition component.

As a method to measure the complexity of nonlinear and non-stationary signals, the sample entropy has the advantages of no need for self-matching and small error [32]. The entropy value represents the complexity of time series data. Therefore, the sample entropy is introduced to evaluate the VMD decomposition effect. The smaller the sample entropy value, the more obvious the periodicity, the less the noise interference, the lower the complexity of the time series data, and the more conducive it is to the training and learning of the gas emission prediction model.

$$SampEn(m, r) = \lim_{N \to \infty} \left\{ -\ln\left[\frac{A^m(r)}{B^m(r)}\right] \right\} \qquad (6)$$

In the formula, m is the window length of the sequence when calculating the sample entropy, r is the similarity tolerance threshold, $A^m(r)$ is the probability of two sequences matching m + 1 points under the similarity tolerance threshold r, and $B^m(r)$ is the probability of two sequences matching m points.

The RMSE and sample entropy are fused to construct the fitness function, which is expressed as Equation (7). It can not only reflect the sequence loss information after decomposition but also contain the sequence decomposition effect.

$$fitness = RMSE \cdot SampEn(m, r) \tag{7}$$

At this point, the VMD parameter selection problem is transformed into the following constrained optimization problem, and the expression is as follows (8):

$$\min_{\alpha, k}\{RMSE * SampEn\}, s.t. \begin{cases} \alpha \in \{200, 2000\} \\ k \in \{3, 10\} \end{cases} \tag{8}$$

Note: k and α optimization range reference [33].

2.3. GA Optimizing VMD

The GA is used to solve the above constraint optimization problem. The steps are as follows:

(1) Input the absolute gas emission time sequence data to be decomposed and set the GA maximum iteration times, population size, crossover probability, and other parameters.

(2) Define the optimization dimension and define its optimization scope.

(3) Initialize the population and generate the initial population. Under the current population, VMD decomposition is performed on the absolute gas emission time series data, and the RMSE and sample entropy of the reconstructed data and the measured data are calculated. The initial best fitness value is calculated, and the initial best chromosome is recorded according to Formula (7).

(4) Iterative optimization is performed according to the maximum number of iterations, and the selection, crossover, and mutation operations are performed to calculate the fitness values of various populations and their respective chromosomes in each iteration.

(5) According to Formula (8), the optimal chromosome is selected and decoded to obtain the optimal values of α and k.

(6) The value of k in the VMD decomposition parameter is a positive integer. The round(k) rounding process is carried out to obtain the final VMD parameter optimization value.

2.4. Sparrow Search Algorithm to Optimize Long and Short-Term Memory Networks' Long Short-Term Memories (LSTM)s

LSTM is a type of deep learning. It solves the problems of gradient explosion and gradient disappearance in RNN training through the "gate" structure [34], which can effectively learn long-term dependence and is widely used in the prediction and classification of time series data. Figure 1 shows the unit structure of LSTM.

Its unit structure realizes information protection and control by forgetting the gate, input gate, and output gate. The LSTM steps are as follows:

(1) Information discard: The cell output at time $t-1$ and cell input at time t are read, and Equation (9) is used to complete the information discarded at time t.

$$f_t = \sigma(W_f \cdot [h_{t-1}, x_t] + b_f) \tag{9}$$

(2) Information update: Equation (10) determines which information needs to be updated by the sigmoid layer, Equation (11) determines how much new information is added to time t by the tanh layer, and the last two parts are combined by Equation (12) to complete the new cell information update.

$$i_t = \sigma(W_i \cdot [h_{t-1}, x_t] + b_i) \tag{10}$$

$$c'_t = \tanh(W_c \cdot [h_{t-1}, x_t] + b_c) \quad (11)$$

$$c_t = f_t \cdot C_{t-1} + i_t \cdot c'_t \quad (12)$$

(3) Information output: Equation (13) is used to determine which part of the cell information needs to be output by the sigmoid layer, the cell state is processed by the tanh layer, and the final information output is completed in conjunction with Equation (14).

$$O_t = \sigma(W_O \cdot [h_{t-1}, x_t] + b_O) \quad (13)$$

$$h_t = O_t \cdot \tanh(C_t) \quad (14)$$

In the formula, h_{t-1} is the output of the previous moment, x_t is the input of the current moment, the sigmoid activation function, tanh is the hyperbolic tangent activation function, W_f, W_i, W_c, and W_O are the weight values of different 'gates', and b_f, b_i, b_c, and b_O are the bias values of different 'gates'.

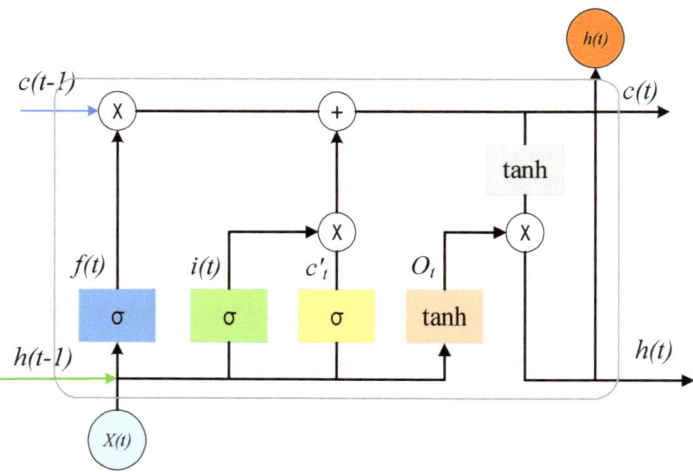

Figure 1. LSTM cell structure.

2.5. SSA Optimizing LSTM

The SSA is a new swarm intelligence optimization algorithm proposed by Xue [35], which is superior to GWO, BA, and other swarm intelligence optimization algorithms in convergence speed, robustness, and stability [36]. In order to avoid over-fitting or under-fitting of the LSTM model caused by human experience, the SSA is introduced to optimize hyperparameters such as MaxEpochs and InitialLearnRate in LSTM so as to establish the optimal gas emission prediction model. The SSA optimization LSTM steps are as follows:

(1) Input the VMD decomposition time sequence data and set the maximum number of iterations of the SSA, population number, security warning value, and other parameters.

(2) Define the optimization dimension and define its optimization scope.

(3) The population is initialized, and the fitness value corresponding to each sparrow is calculated according to Equation (15) and sorted. The initial global optimal fitness value is determined according to the sorting result, and the initial global optimal position is recorded.

$$fitness = MSE = \frac{1}{n}\sum_{i=1}^{n}(IMF - IMF') \quad (15)$$

In the formula, n is the number of *IMF* samples, *IMF* is the modal data of the VMD decomposition, and *IMF'* is the prediction data of the LSTM model.

(4) Iterative optimization is performed according to the maximum number of iterations, and the fitness values corresponding to each sparrow under each iteration are calculated and their positions recorded.

(5) Optimize the best fitness value and the best position according to Equation (16), and the best position obtained is the final optimization value of each parameter.

$$\min\{MSE\}, s.t. \begin{cases} \text{numHiddenUnits} \in \{2, 200\} \\ \text{InitialLearnRate} \in \{0.0001, 1\} \\ \text{L2Regularization} \in \{0.00001, 1\} \\ \text{MaxEpochs} \in \{2, 300\} \end{cases} \quad (16)$$

2.6. Construction of GA-VMD-LSTM Prediction Model

Based on the above analysis, the process of the prediction model in Figure 2 is constructed. The specific steps are as follows:

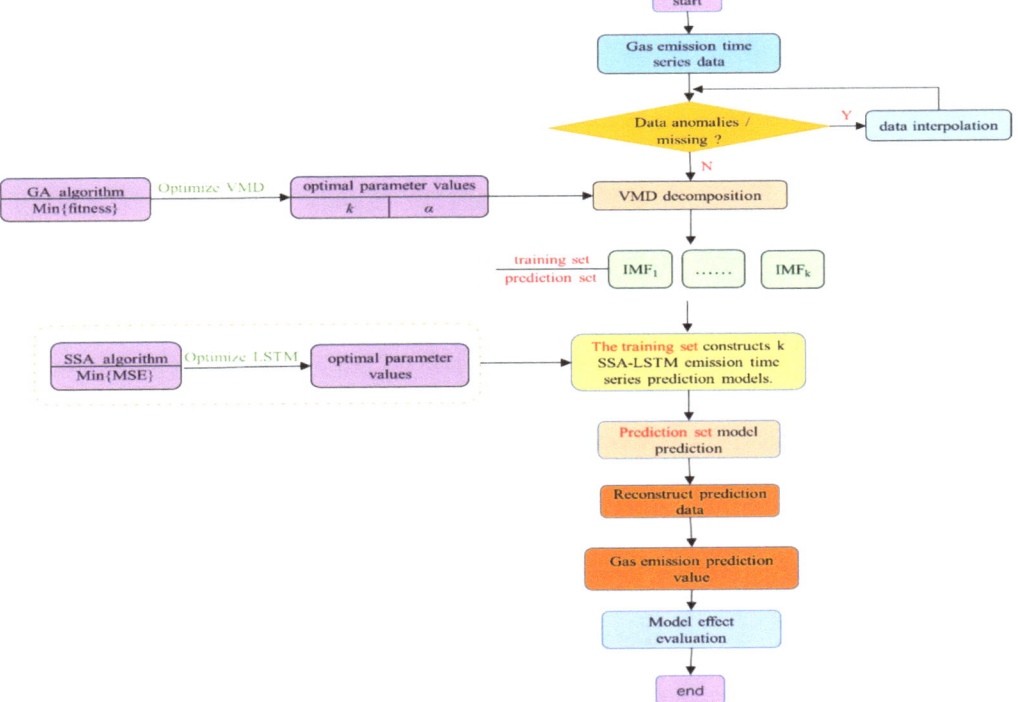

Figure 2. Flow diagram of gas emission prediction model.

(1) Pre-processing of the absolute gas emission time series data. Detect missing values and outliers to ensure data integrity.

(2) VMD data decomposition. Firstly, the new fitness was constructed as the evaluation standard, and then the GA optimization algorithm was used to optimize the VMD parameters k and α to obtain the optimal VMD parameter settings. In this way, the timing data are decomposed into IMF1, IMF2, ..., IMFk.

(3) SSA-LSTM model prediction. The IMFk decomposition data are divided into a training set and a prediction set. The training set data are used to build the SSA-LSTM

model, and the prediction set data are used to predict the SSA-LSTM model and output the predicted value.

(4) According to the principle of equal weight superposition, the predicted values of each model were superimposed to obtain the final prediction results of the gas emission.

(5) Model effect evaluation. The mean absolute error (*MAE*), mean absolute percentage error (*MAPE*), root mean square error (*RMSE*), and decision coefficient (R2) were used to evaluate the effect of the prediction model. The equation is from (17) to (20):

$$MAE = \frac{1}{n}\sum_{i=1}^{n}|\alpha_t - \hat{\alpha}_t| \quad (17)$$

$$MAPE = \frac{1}{n}\sum_{i=1}^{n}\left|\frac{\alpha_t - \hat{\alpha}_t}{\alpha_t}\right| \quad (18)$$

$$RMSE = \sqrt{\frac{1}{n}\sum_{i=1}^{n}(\alpha_t - \hat{\alpha}_t)^2} \quad (19)$$

$$R^2 = 1 - \frac{\sum_{i=1}^{n}(\alpha_i - \hat{\alpha}_i)^2}{\sum_{i=1}^{n}(\alpha_i - \overline{\alpha_i})^2} \quad (20)$$

In the equation, α_i is the measured data of the gas emission, $\hat{\alpha}_i$ is the predicted data of the gas emission, and n is the number of samples collected.

3. Results and Discussion

Taking the 1417 fully mechanized mining face of a coal mine in Shaanxi as the research object, the main coal seam of the 1417 working face is a 4-2 coal seam; the thickness of the coal seam is 4.0~19.0 m, and the average thickness is 10.0 m. The working face adopts the gas control measures of "pre-mining strata hole pre-extraction + roof directional long drilling + upper corner buried pipe extraction + air exhaust".

The absolute gas emission data of the 1417 working face, from 26 January 2022 to 30 April 2022, were collected, as shown in Table 1.

Table 1. Gas emission data.

Serial Number	Time	Class	Gas Ventilation Volume/ (m³·min⁻¹)	Gas Drainage Volume/ (m³·min⁻¹)	Absolute Gas Emission Quantity/ (m³·min⁻¹)
1	1/26	16	5.25	10.91	16.16
2	1/27	0	5.51	12.17	17.68
3	1/27	8	4.99	12.70	17.69
4	1/27	16	5.51	12.10	17.61
5	1/28	0	6.30	14.59	20.89
……	……	……	……	……	……
280	4/29	8	3.15	11.44	14.59
281	4/29	16	3.15	10.93	14.08
282	4/30	0	3.38	11.35	14.73
284	4/30	8	2.93	10.99	13.92
283	4/30	16	4.05	11.81	15.86

The outliers and missing values should be detected before constructing the time series prediction model. Data points other than ±1.5 IRQ (the IQR represents the interquartile distance) were taken as outliers to draw a box plot in Figure 3.

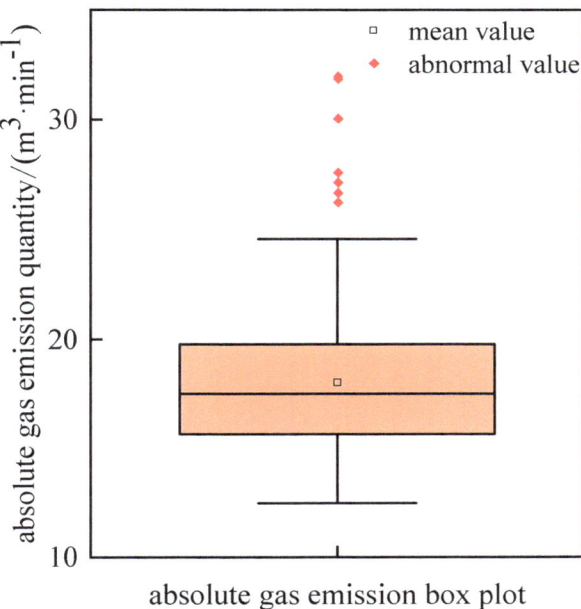

Figure 3. Outlier discriminant boxplot.

It can be seen from Figure 3 that there are seven outliers in the collected data, and the specific values are shown in Table 2.

Table 2. Outliers of gas emission data.

Serial Number	Time	Class	Absolute Gas Emission Quantity/($m^3 \cdot min^{-1}$)
23	2/3	0	23.22
61	2/15	16	26.66
62	2/16	0	27.14
64	2/16	16	31.95
65	2/17	0	31.85
72	2/19	8	27.59
75	2/20	8	30.04

For missing data, through the statement shuju [!complete.cases (shuju),], data were detected and no missing values were found.

3.1. Data INTERPOLATION

The Table 2 outliers were deleted and interpolated. In order to obtain the best filling method, the outflow data without outliers (99 groups of data from 21 February 2022 to 25 March 2022) were extracted from the original data for random missing [37] processing. The EM interpolation, mean interpolation, linear interpolation, and random forest interpolation were used for interpolation processing, and their mean square errors were compared to optimize the best interpolation method. The mean square error of each interpolation method is shown in Table 3.

Table 3. Imputation error comparison for random missing.

Absence Rate/%	Mean Square Error of Different Interpolation Methods			
	EM Algorithm Imputation	Mean Imputation	linear Interpolation	Random Forest Imputation
5	2.30	3.04	0.11	3.47
10	1.28	1.78	0.13	1.81
15	1.13	1.42	0.16	1.48
sor	1.53	1.66	0.39	1.84
25	1.48	1.81	0.41	2.00
30	1.62	2.20	0.39	2.14

As can be seen in Table 3, linear interpolation has the highest interpolation accuracy under six types of miss rates, so linear interpolation is selected for interpolation. The linear interpolation data are shown in Table 4.

Table 4. Linear interpolation fill data.

Serial Number	Time	Class	Absolute Gas Emission Quantity/($m^3 \cdot min^{-1}$)
23	2/3	0	21.41
61	2/15	16	23.99
62	2/16	0	23.50
64	2/16	16	22.46
65	2/17	0	21.92
72	2/19	8	24.00
75	2/20	8	23.73

3.2. VMD Decomposition of Gas Emission Data

The time series data of the absolute gas emission at the 1417 working face after linear interpolation, from 26 January 2022 to 30 April 2022, totaled 283 sets of sample data. The timing diagram is shown in Figure 4

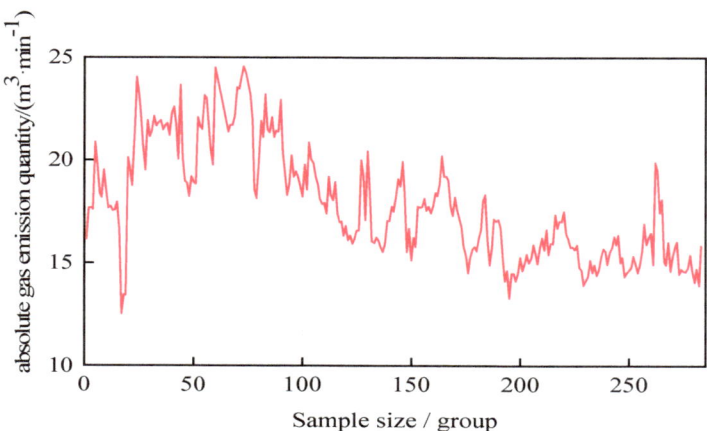

Figure 4. The 1417 mining working face timing diagram.

Firstly, the GA algorithm was used to optimize the VMD parameters. The GA-related parameter settings are as follows: the maximum number of iterations is 10; population size 10; crossover probability 0.8; variation probability 0.1; k optimization range [3, 10]; α optimization range [200, 2000]. Its iterative optimization curve is shown in Figure 5.

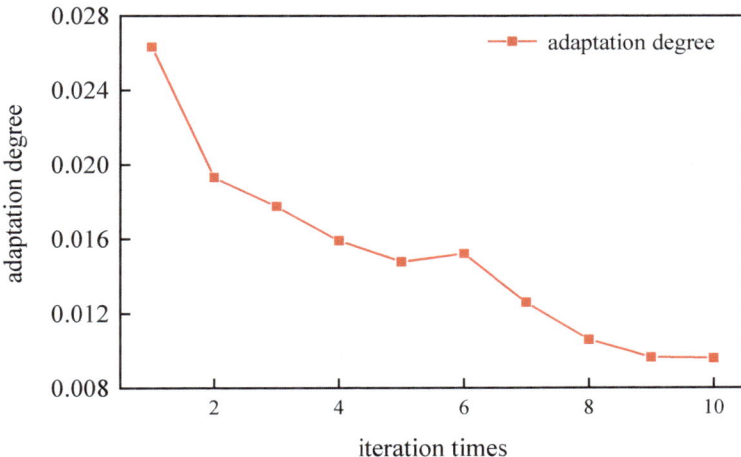

Figure 5. Iterative optimization curve of GA to optimize VMD.

As can be seen in Figure 5, at the tenth iteration, the minimum fitness value of 0.0096 is obtained, k is 10, and α is 483.70. The VMD decomposition results are shown in Figure 6.

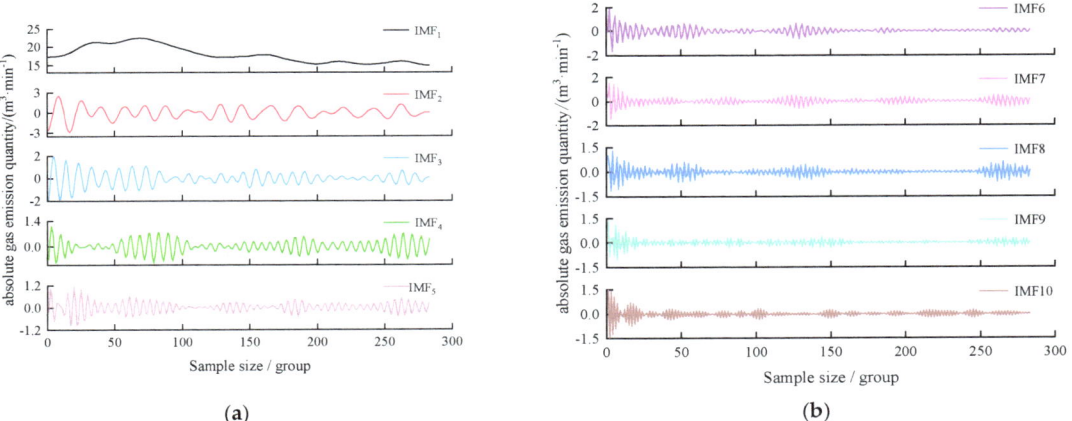

Figure 6. Gas emission of the No. 1417 mining working face by VMD (**a**) IMF1~IMF5 (**b**) IMF6~IMF10.

As can be seen in Figure 6, 10 IMF decomposition components of different frequencies were obtained after the VMD decomposition of the absolute gas emission time series data. The IMF1 component, which characterizes the trend change for the absolute gas emission, and the IMF2~IMF10 components with certain periodic characteristics are obtained by decomposition, which reduces the complexity of the original data. In order to evaluate the optimal VMD decomposition effect after GA optimization (k = 10, α = 483.70), the decomposition results of k = 3, 5, 8 (α = 2000) were compared with those of k = 10, and the comparison results are shown in Figure 7 and Table 5.

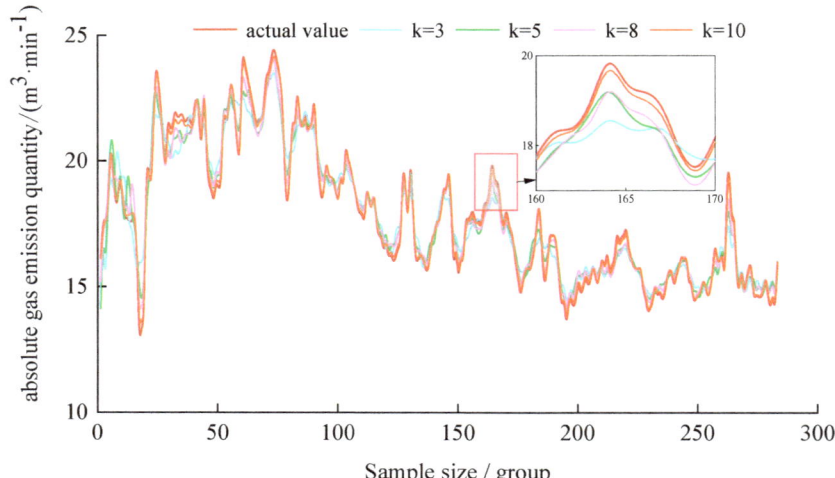

Figure 7. Reconstruction curve by VMD.

Table 5. Reconstruction data by VMD.

Serial Number	Actual Value ($m^3 \cdot min^{-1}$)	VMD Decomposition Reconstruction Value ($m^3 \cdot min^{-1}$)			
		k = 3	k = 5	k = 8	k = 10
1	16.16	16.16	14.13	15.18	16.10
2	17.68	17.68	17.27	17.26	17.52
3	17.69	17.69	15.60	17.61	17.56
4	17.61	17.61	18.91	18.48	17.87
5	20.89	20.89	20.78	20.64	20.87
......	......	......	......	......	......
280	14.59	14.69	14.92	14.79	14.66
281	14.08	15.23	14.85	14.77	14.17
282	14.73	15.29	15.45	15.28	14.87
284	13.92	14.64	14.72	14.32	14.07
283	15.86	15.37	15.53	16.05	16.04

According to Figure 7, when k = 10, the coincidence between the reconstructed value curve of VMD and the actual value curve of the absolute gas emission is the best. At the sudden change point of gas emission, it is also the closest to the original data. While reducing the complexity of the data, the fluctuation information of the original data is retained. The RMSE values of k = 3, 5, 8, and 10 calculated from Table 5 are 0.87, 0.66, 0.46, and 0.13, respectively, and the data decomposition loss of k = 10 is the lowest. The VMD decomposition effect optimized by the GA is superior to the VMD decomposition effect set by the empirical value in decomposition loss and mutation data retention.

3.3. Prediction of Gas Emission

In order to verify the effectiveness of the VMD decomposition algorithm, the decomposed data were divided into training sets and prediction sets in a 4:1 ratio.

The training set data realizes the optimization of the key parameters of LSTM by the SSA, and the optimization value is shown in the table.

The optimal LSTM model parameters were determined by the SSA optimization values in Table 6, thus completing the construction of the GVSL prediction model.

Table 6. LSTM parameter values for SSA optimization.

Decomposed Component	Num Hidden Units	Max Epochs	InitialLearnRate	L2 Regularization
IMF1	161	255	0.0319	0.0284
IMF2	98	54	0.0631	0.0604
IMF3	200	81	0.0081	0.0001
IMF4	21	16	0.7578	0.8235
IMF5	36	30	0.1323	0.1069
IMF6	115	25	0.0118	0.0001
IMF7	30	73	0.0001	0.0001
IMF8	12	25	0.0001	0.0001
IMF9	6	9	0.0011	0.0010
IMF10	200	60	0.0116	0.0001

In order to verify the prediction effect of the model, the GVSL prediction model is used to predict the absolute gas emissions of 57 groups in the future of the prediction set. The prediction results are shown in Figure 8, and the prediction errors are shown in Table 7.

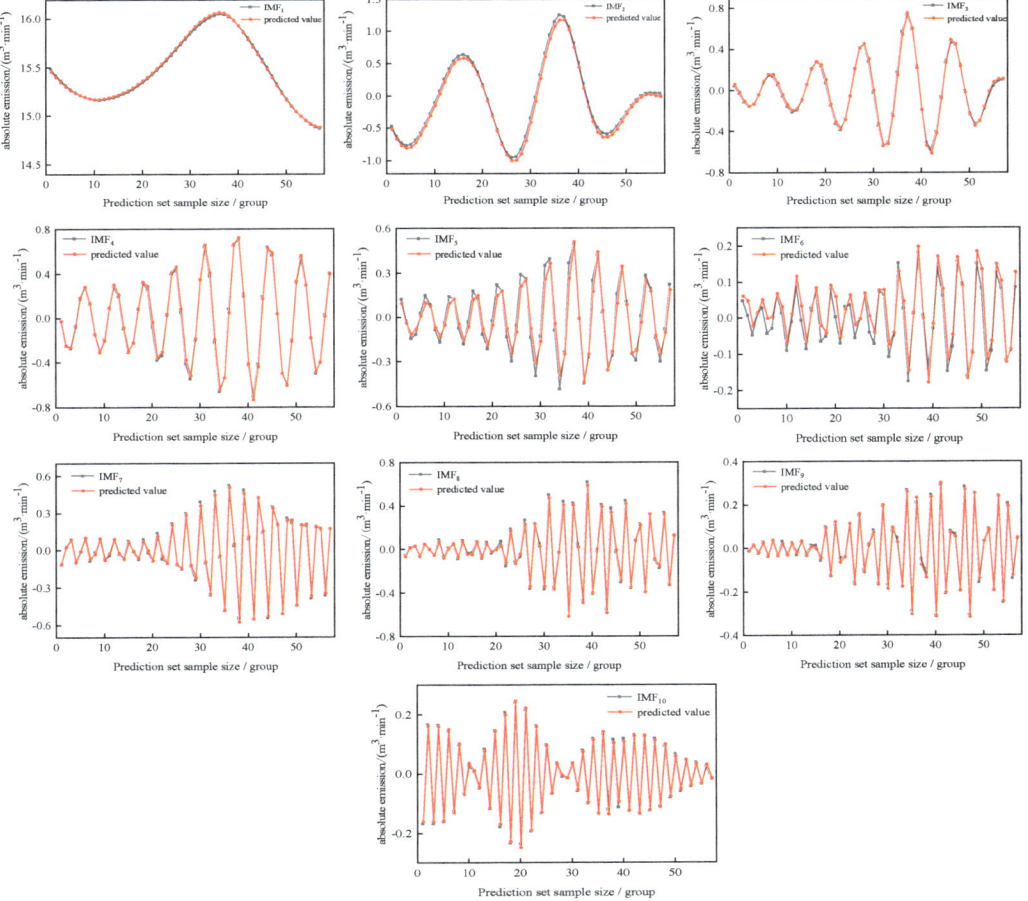

Figure 8. The prediction results of each decomposition component by GVSL.

Table 7. Prediction error of each GVSL model.

GVSL Forecasting Model	Absolute Error (m$^3 \cdot$min^{-1})		
	Minimum Value	Maximum Value	Mean Value
IMF1	0.0009	0.0266	0.0152
IMF2	0.0178	0.0808	0.0460
IMF3	0.0003	0.0426	0.0146
IMF4	0.0002	0.0648	0.0140
IMF5	0	0.0999	0.0348
IMF6	0	0.0583	0.0218
IMF7	0.0005	0.0307	0.0109
IMF8	0.0006	0.0399	0.0144
IMF9	0.0002	0.0254	0.0083
IMF10	0.0010	0.0189	0.0047

Note: To avoid multiple zeros in two decimal places, four decimal places are reserved here.

It can be seen in Figure 8 and Table 7 that the fitting degree of the predicted value curve and the actual value curve of each component of the GVSL prediction model is high. The average absolute error of prediction of each model fluctuated in the range of 0.0047~0.0460 m^3/min and was maintained at a low level. The GVSL prediction model of each component has a better prediction effect and successfully predicts the changing trend of each VMD decomposition component.

The prediction results of each component of the GVSL model were superimposed to obtain the final prediction results of absolute gas emission, as shown in Figure 9.

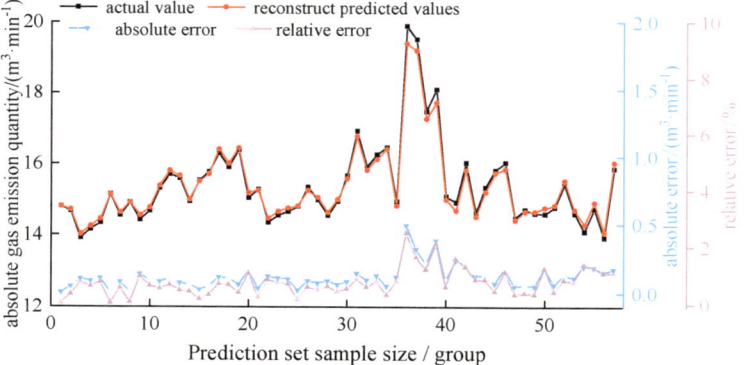

Figure 9. Reconstructed predictions by GVSL model.

It can be seen from Figure 9 that the reconstructed value curve obtained by each GVSL model coincides with the actual value curve of absolute gas emission. The absolute error ranges from 0.0014 to 0.4895 m^3/min, and the average absolute error is 0.1156 m^3/min. The relative error ranges from 0.01% to 2.46%, and the average absolute error is 0.73%. The model can well predict the trend of absolute gas emission in the prediction of the 57 sets.

3.4. Comparative Analysis of Prediction Models

In the three scenarios (scenario 1: the sample size of the training set was 162, and the sample size of the prediction set was 121; scenario 2: the sample size of the training set was 189 groups, and that of the prediction set was 94 groups; scenario 3: the training set sample size was 226, and the prediction set sample size was 57), we compared and analyzed the prediction effects of the GVSL, VMD-LSTM, SSA-LSTM, and GPR models. The results are shown in Figure 10 and Table 8.

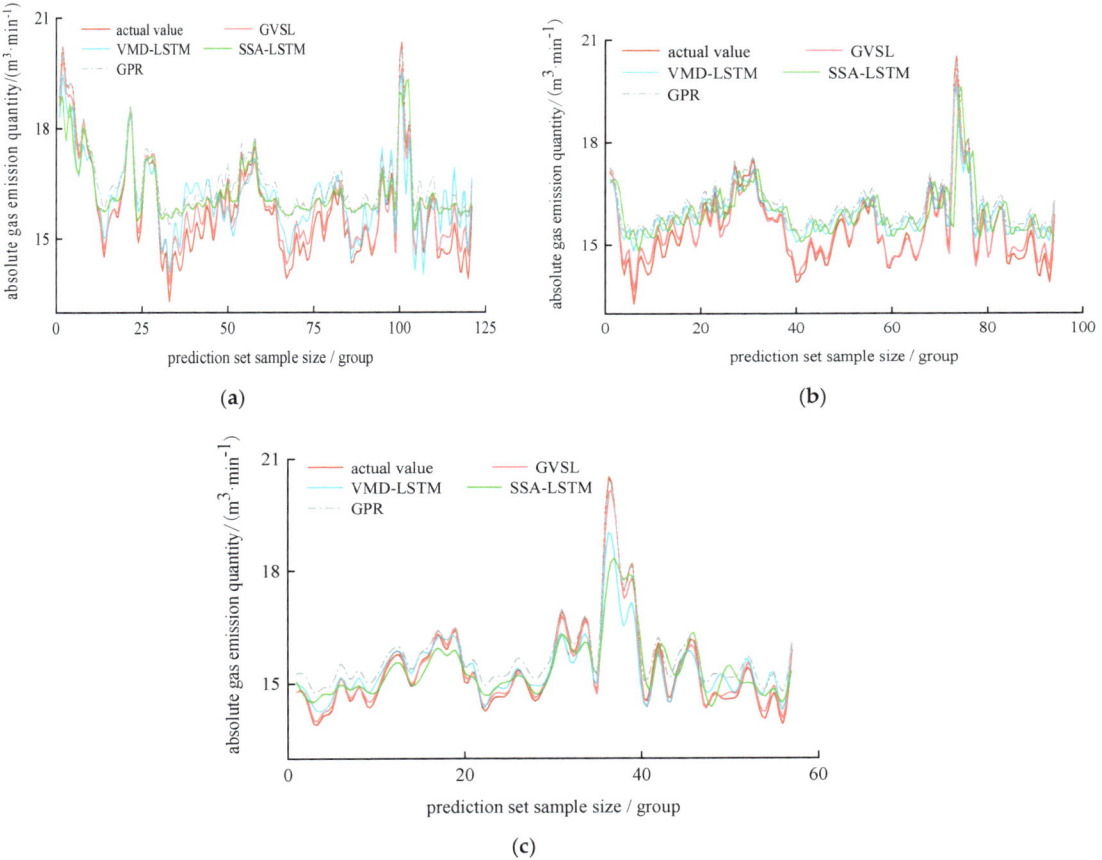

Figure 10. Comparison of different prediction models. (**a**) scenario 1 (**b**) scenario 2 (**c**) scenario 3.

Table 8. Model evaluation index comparison.

	Evaluating Indicator	GVSL	VMD-LSTM	SSA-LSTM	GPR
Scenario one	MAE	0.27	0.60	0.65	0.77
	MAPE/%	1.72	3.82	4.27	5.14
	RMSE	0.31	0.72	0.83	0.88
	R2	0.95	0.74	0.67	0.68
Scenario two	MAE	0.18	0.52	0.73	0.68
	MAPE/%	1.16	3.50	4.76	4.56
	RMSE	0.22	0.61	0.97	0.77
	R2	0.96	0.74	0.34	0.65
Scenario three	MAE	0.11	0.30	0.37	0.39
	MAPE/%	0.71	1.91	2.33	2.62
	RMSE	0.14	0.41	0.53	0.44
	R2	0.99	0.88	0.80	0.87

It can be seen in Figure 10 and Table 8 that the GVSL model has the best prediction effect, especially at the sudden change point of gas emission. It can be clearly seen that the GVSL model is superior to other models, which verifies the feasibility of the model for predicting the absolute gas emission of the working face. The MAE, MAPE, RMSE, and R2 values of the GVSL, VMD-LSTM, SSA-LSTM, and GPR models are compared horizontally. It is concluded that the prediction effect of the GVSL model is better than that of the other three models.

The MAE, MAPE, RMSE, and R2 values of the GVSL model in scenario 1, scenario 2, and scenario 3 were compared, and the R2 values were 0.95, 0.96, and 0.99, respectively. The R2 of the GVSL model in scenario 3 is 0.99, which is closer to 1; that is, the larger the proportion of the training set and prediction set, the more advantages the GVSL model has.

4. Conclusions

(1) The interpolation accuracy of random forest interpolation, mean interpolation, EM interpolation, and linear interpolation is compared and analyzed under six types of missing rates. It is determined that linear interpolation is used to interpolate the outliers to ensure the integrity of the data structure.

(2) The optimal k value and α value of VMD under the new fitness function are 10 and 483.70. Through comparative analysis of the VMD decomposition results when k = 3, 5, 8, and 10, it was found that the data decomposition loss of k = 10 is 0.13, which is lower than the pair ratio, and the VMD decomposition effect after GA optimization is the best.

(3) The prediction effects of the GVSL, VMD-LSTM, SSA-LSTM, and GPR models were compared and analyzed under three scenarios. The results show that the prediction accuracy of the GVSL model is the highest, which proves that the model can be effectively applied to the prediction of gas emission in the coal face.

Author Contributions: Conceptualization, Z.Y. and J.Z. (Jingzhao Zhang); methodology, Y.C.; software, Y.C.; validation, Y.C., Y.H. and C.Z.; formal analysis, Y.C.; investigation, J.G.; resources, J.Z. (Jingzhao Zhang); data curation, Y.C.; writing—original draft preparation, Y.C.; writing—review and editing, Z.Y.; visualization, Y.H.; supervision, J.Z. (Jinlong Zhang); project administration, Z.Y.; funding acquisition, F.Z. All authors have read and agreed to the published version of the manuscript.

Funding: This research was funded by the Key Research and Development Program of Shaanxi Province, grant numbers 2020GY-139 and 2022GY-150.

Institutional Review Board Statement: Not applicable.

Informed Consent Statement: Not applicable.

Data Availability Statement: The data presented in this study are available upon request from the corresponding author due to privacy.

Conflicts of Interest: The authors declare no conflicts of interest.

References

1. Fu, H.; Yu, X.; LU, W. Based on Ant Colony and Particle Swarm Hybrid Algorithm and LS-SVM Gas Emission Prediction. *J. Transduct. Technol.* **2016**, *29*, 373–377.
2. Huang, W.; Tong, M.; Ren, Z. Nonlinear Combination Prediction Method of Gas Emission Based on SVM. *J. China Univ. Min. Technol.* **2009**, *38*, 234–239.
3. Fu, H.; Xie, S.; Xu, Y.; Chen, Z. Research on dynamic prediction model of coal mine gas emission based on ACC-ENN algorithm. *J. China Coal Soc.* **2014**, *39*, 1296–1301.
4. Zhang, L.; Qin, Y.; Jiang, W.; Jing, H.; Zhao, G. Research status and prospects of mine gas emission prediction methods in my country. *Saf. Coal Mines* **2007**, *38*, 58–60.
5. Fan, B.; Bai, C.; Li, J. Prediction of gas emission from coal mining face based on LMD-SVM. *J. Min. Saf. Eng.* **2013**, *30*, 946–952.
6. Fu, H.; Xie, S.; Xu, Y.; Chen, Z. Research on prediction model of mine gas emission based on MPSO-WLS-SVM. *Chin. Saf. Sci. J.* **2013**, *23*, 56–61.
7. Dong, X.; Jia, J.; Bai, Y.; Fan, C. Prediction of gas emission from coal mining face based on SVM coupled genetic algorithm. *J. Saf. Environ.* **2016**, *16*, 114–118.
8. Feng, S.; Shao, L.; LU, W.; Meng, T.; Gao, Z. Application of PCA-PSO-LSSVM Model in Gas Emission Prediction. *J. Liaoning Technol. Univ. (Nat. Sci.)* **2019**, *38*, 124–129.
9. Wang, Y. Gas emission prediction based on PCA-PSO-ELM. *J. Hunan Univ. Sci. Technol. (Nat. Sci. Ed.)* **2020**, *35*, 1–9.

10. Zhang, Q.; Jia, B.; Dong, X.; Li, Z. Prediction of gas emission in mining face by PCA-GA-SVM. *J. Liaoning Technol. Univ. (Nat. Sci.)* **2015**, *34*, 572–577.
11. Xiang, P.; Xie, X.; Shuang, H.; Liu, C.; Wang, H.; Xu, J. Research on Gas Emission Prediction Based on KPCA-CMGANN Algorithm. *China Saf. Sci. J.* **2020**, *30*, 39–47.
12. Wang, L.; Liu, Y.; Liu, Z.; Qi, J. Research on Gas Emission Prediction Model Based on IABC-LSSVM. *Sens. Microsyst.* **2022**, *41*, 34–38.
13. Huang, W.; Shi, S. Gas gushing time series prediction based on improved Lyapunov index. *J. China Coal Soc.* **2009**, *34*, 1665–1668.
14. Dan, Y.; Hou, F.; Fu, H.; Ma, J. Prediction of Gas Emission in Chaotic Time Series Based on Improved Extreme Learning Machine. *China Saf. Sci. J.* **2012**, *22*, 58–63.
15. Liu, J.; An, F.; Lin, D.; Guo, Z.; Zhang, L. Natural mode SVM modeling and prediction of gas emission from coal mining face. *Syst. Eng. Theory Pract.* **2013**, *33*, 505–511.
16. Lu, G.; Li, X.; Zu, B.; Dong, J. Research on time-varying sequence prediction of gas emission based on EMD-MFOA-ELM. *J. Saf. Sci. Technol.* **2017**, *13*, 109–114.
17. Dai, W.; Fu, H.; Ji, C. VMD-DE-RVM interval prediction method for gas emission in mining face. *China Saf. Sci. J.* **2018**, *28*, 109–115.
18. Xiao, P.; Xie, H.; Shuang, H.; Liu, C.; Xu, J.; Hong, J. Application of Wavelet-Extreme Learning Machine in Time-varying Sequence Prediction of Gas Emission. *J. Xi'an Univ. Sci. Technol.* **2020**, *40*, 839–845.
19. Zhan, G.; Wang, Y.; Fu, H.; Wang, S. Prediction of Gas Emission Based on Variational Mode Decomposition and Deep Integration Combination Model. *Control Eng. China* **2022**, *29*, 1–12.
20. Dragomiretskiy, K.; Zosso, D. Variational Mode Decomposition. *IEEE Trans. Signal Process.* **2014**, *62*, 531–544. [CrossRef]
21. Zhang, Y.; Han, P.; Wang, D.; Wang, S. Short-term wind speed prediction of wind farms based on variational mode decomposition and LSSVM. *Acta Energiae Solaris Sin.* **2018**, *39*, 194–202.
22. Zhang, S.; Su, X.; Chen, R.; Liu, W.; Zuo, Y.; Zhang, Q. Short-term power load forecasting based on variational modal decomposition and FABP. *Chin. J. Sci. Instrum.* **2018**, *39*, 67–73.
23. Balakrishnan, R.; Geetha, V.; Kumar, M.R.; Leung, M.-F.; Lucchi, E. Reduction in Residential Electricity Bill and Carbon Dioxide Emission through Renewable Energy Integration Using an Adaptive Feed-Forward Neural Network System and MPPT Technique. *Sustainability* **2023**, *15*, 14088. [CrossRef]
24. Ma, H.; Tong, Q.; Zhang, Y. Application of Variational Mode Decomposition of Optimization Parameters in Fault Diagnosis of Rolling Bearings. *China Mech. Eng.* **2018**, *29*, 390–397.
25. Zhang, S.; Li, J.; Jiang, A.; Huang, J.; Liu, W.; Ai, H. Novel two-stage short-term power load forecasting based on FPA-VMD and BiLSTM neural network. *Power Syst. Technol.* **2022**, *46*, 3269–3279.
26. LeCun, Y.; Bengio, Y.; Hinton, G. Deep learning. *Nature* **2015**, *521*, 436–444. [CrossRef]
27. Liu, Y.; Zhao, Q. CNN-LSTM ultra-short-term power load forecasting based on cluster empirical mode decomposition. *Power Syst. Technol.* **2021**, *45*, 4444–4451.
28. Meng, A.; Xu, X.; Chen, J.; Wang, C.; Zhou, T.; Yin, H. Ultra-short-term photovoltaic power prediction based on reinforcement learning and combined deep learning model. *Power Syst. Technol.* **2021**, *45*, 4721–4728.
29. Hao, S.; He, T.; Ma, X.; Zhang, X.; Wu, Y.; Wang, H. KDBiDet: A Bi-Branch Collaborative Training Algorithm Based on Knowledge Distillation for Photovoltaic Hot-Spot Detection Systems. *IEEE Trans. Instrum. Meas.* **2024**, *73*, 3504615. [CrossRef]
30. Liu, J.; Quan, H.; Yu, X.; Li, Z. Fault diagnosis of rolling bearing based on parameter optimization VMD and sample entropy. *Acta Autom. Sin.* **2022**, *48*, 808–819.
31. Chen, C.; Li, X.; Yang, L.; Qu, H.; Wang, Y.; He, C. Application of Variational Mode Decomposition in Power System Harmonic Detection. *Power Syst. Prot. Control* **2018**, *46*, 63–70.
32. Yang, D.; Feng, F.; Zhao, Y.; Jiang, P.; Ding, C. VMD sample entropy feature extraction method and its application in planetary gearbox fault diagnosis. *J. Vib. Shock* **2018**, *37*, 198–205.
33. Zhao, X.; Zhang, S.; Li, Z.; Li, F.; Hu, Y. Fault feature signal extraction method based on VMD. *J. Vib. Meas. Diagn.* **2018**, *38*, 11–19, 202.
34. Hochreiter, S.; Schmidhuber, J. Long short-term memory. *Neural Comput.* **1997**, *9*, 1735–1780. [CrossRef]
35. Li, Y.; Wang, S.; Chen, Q.; Wang, X. Comparative Research on Several New Swarm Intelligence Optimization Algorithms. *Comput. Eng. Appl.* **2020**, *56*, 1–12. [CrossRef]
36. Zhao, H.; Shen, X.; Lv, L.; Lan, P.; Liu, J.; Liu, D. Load data restoration based on GAN and its application in short-term load forecasting of EV. *Autom. Electr. Power Syst.* **2021**, *45*, 143–151.
37. Ma, Z.; Li, Y.; Liu, Z.; Gu, C. Fault Feature Extraction of Rolling Bearing Based on Variational Mode Decomposition and Teager Energy Operator. *J. Vib. Shock* **2016**, *35*, 134–139.

Disclaimer/Publisher's Note: The statements, opinions and data contained in all publications are solely those of the individual author(s) and contributor(s) and not of MDPI and/or the editor(s). MDPI and/or the editor(s) disclaim responsibility for any injury to people or property resulting from any ideas, methods, instructions or products referred to in the content.

Article

Indeterministic Data Collection in UAV-Assisted Wide and Sparse Wireless Sensor Network

Yu Du [1], Jianjun Hao [2], Zijing Chen [2] and Yijun Guo [2,*]

[1] Business School, Beijing Language and Culture University, Beijing 100083, China; duyu@blcu.edu.cn
[2] Beijing Key Laboratory of Network System Architecture and Convergence, Beijing University of Posts and Telecommunications, Beijing 100876, China; jjhao@bupt.edu.cn (J.H.); chenzijing@bupt.edu.cn (Z.C.)
* Correspondence: guoyijun@bupt.edu.cn

Abstract: The widespread adoption of Internet of Things (IoT) applications has driven the demand for obtaining sensor data. Using unmanned aerial vehicles (UAVs) to collect sensor data is an effective means in scenarios with no ground communication facilities. In this paper, we innovatively consider an indeterministic data collection task in a UAV-assisted wide and sparse wireless sensor network, where the wireless sensor nodes (SNs) obtain effective data randomly, and the UAV has no pre-knowledge about which sensor has effective data. The UAV trajectories, SN serve scheduling and UAV-SN association are jointly optimized to maximize the amount of collected effective sensing data. We model the optimization problem and address the indeterministic effective indicator by introducing an effectiveness probability prediction model. The reformulated problem remains challenging to solve due to the number of constraints varying with the variable, i.e., the serve scheduling strategy. To tackle this issue, we propose a two-layer modified knapsack algorithm, within which a feasibility problem is resolved iteratively to find the optimal packing strategy. Numerical results demonstrate that the proposed scheme has remarkable advantages in the sum of effective data blocks, reducing the completion time for collecting the same ratio of effective data by nearly 30%.

Keywords: trajectory planning; indeterministic data collection; wireless sensor network; internet of things

Citation: Du, Y.; Hao, J.; Chen, Z.; Guo, Y. Indeterministic Data Collection in UAV-Assisted Wide and Sparse Wireless Sensor Network. *Sensors* **2024**, *24*, 6496. https://doi.org/10.3390/s24196496

Academic Editors: Chunhui Zhao and Shuai Hao

Received: 14 September 2024
Revised: 2 October 2024
Accepted: 6 October 2024
Published: 9 October 2024

Copyright: © 2024 by the authors. Licensee MDPI, Basel, Switzerland. This article is an open access article distributed under the terms and conditions of the Creative Commons Attribution (CC BY) license (https://creativecommons.org/licenses/by/4.0/).

1. Introduction

1.1. Background and Motivation

With the rapid development of the Internet of Things (IoT), wireless sensor networks (WSNs) have been widely applied in many consumer electronics applications, such as intelligent transportation, forest monitoring, smart farms, smart ocean, E-commerce, environmental monitoring and emergency rescue [1,2]. Many studies have been conducted in this field. Refs. [3,4] studied the physical layer techniques, Refs. [5–7] focused on optimizing the network efficiency, Refs. [8,9] tackled the problem of network security, and [10,11] researched on the analysis and mining of sensing data. A basic and key problem faced by these smart applications is acquiring sensing data from sensor nodes timely and effectively. It is predicted that the number of sensors in the world is expected to exceed 100 trillion by 2030 [12]. Hence, in WSNs, how to collect a huge amount of sensor data while satisfying the low-energy consumption, low delay, and high reliability requirements of IoT applications is a challenging problem.

In many IoT applications, wireless sensor nodes (SNs) have been deployed in remote and harsh environments [13], where it is inconvenient to deploy ground infrastructure to collect sensing data from SNs. Due to the advantages of low-cost, small size, flexibility and high mobility, UAVs have recently been employed in data collection for WSNs, which leads to the so-called UAV-assisted data collection network [14]. Furthermore, UAVs can fly near the wireless sensor nodes to achieve highly energy-efficient data transmissions over the line-of-sight (LoS) communication links, which is very helpful in lowering the energy consumption for SNs to transmit data, and hence extending the survival time of WSNs.

In UAV-assisted data collection networks, UAVs have found a substantial performance improvement with respect to effectiveness indicators, such as communication capacity and max-min rate, via jointly optimizing UAV trajectories and UAV-SN associations [15,16]. To further improve the spectral efficiency and support massive connectivity, NOMA is integrated with its user grouping and power allocation being optimized jointly to maximize the sum rate of a wireless sensor network [17]. The above works regarding UAV trajectory designs have assumed a fixed operation period for UAVs. In order to be compatible with time-sensitive services, the task completion time is minimized in cases of constrained energy [18] and NOMA-enabled [19]. In [20], the authors analytically characterize the optimal solution structure for the joint UAV trajectory design and SN scheduling. In [21], the energy budget of ground sensors is taken into account to further lower UAVs' completion time. By taking the energy limitation of WSNs into consideration, a joint 3D trajectory design and data collection scheduling scheme is encouraged to save the energy of both UAVs and SNs [22], and the long-term energy consumption is minimized in [23]. Moreover, for large-scale IoT where a large amount of sensors are deployed, clustering sensors can significantly improve data acquisition efficiency. A cluster head selection scheme and the corresponding data forwarding rules within the cluster are proposed to maximize the value of information (VoI) [24]. Considering a multi-scenario parallel data collection task, the clustering strategy, cluster head mode selection, UAV flight trajectory and UAV velocity are jointly optimized to minimize the data collection time [25]. In [26], the clustering algorithm is improved to optimize the system's energy efficiency. Besides, for applications that require real-time updates, such as connected vehicle networks, remote monitoring systems, etc., the AoI-minimal data collection is considered for UAV-assisted WPCNs [27] as well as UAV-aided IoT networks [28].

1.2. Contribution

It should be emphasized that all the aforementioned studies have focused on deterministic data collection tasks. In these scenarios, the locations of SNs and the fact that each SN has acquired valid data for collection are both predetermined and known in advance. However, In many IoT applications, such as wildlife monitoring and forest fire detection, ground sensors are sparsely distributed over vast areas. While these SNs continuously monitor their surroundings, they only intermittently capture meaningful data, like footage of wildlife activities, that require collection. In other words, the occurrence of SNs obtaining valuable data is probabilistic. Therefore, for a wide and sparse WSN, and considering the limited battery life on UAVs, maximizing the collection of valid data within the constraints of a UAV's flight time presents a significant challenge. To the best of our knowledge, no studies have yet addressed the issue of indeterministic data collection.

To address the indeterministic data collection problem mentioned above, we first present a model of UAV-assisted wide and sparse wireless sensor network (WS-WSN), and formulate an effective data block sum maximization problem. Then, we develop a joint UAV trajectories design, SN serve scheduling and UAV-SN association algorithm is developed. Numerical results show that our proposed algorithm has better performance in the ratio of collected effective data blocks compared to baseline algorithms. The contributions presented in this paper are summarized as follows:

- We model a UAV-assisted wide and sparse wireless sensor network and formulate a novel indeterministic data collection problem. By our consideration, only a part of ground SNs obtain effective sensing data that contain target information. Under the wide and sparse wireless sensor network assumption, UAVs are not able to fly over and serve all of the SNs ergodically due to limited onboard energy. Accordingly, we formulate an effective data block sum maximization problem that aims to maximize the number of effective data blocks within a limited flying period.

- We propose a joint UAV trajectories design, SN serve scheduling and UAV-SN association algorithm. In particular, to deal with the indeterministic effectiveness indicator, we reformulate the problem by introducing an effective probability prediction model based on Deep Neural Network (DNN). Furthermore, to tackle the difficulty of varying constraints brought by partial data collection, a modified knapsack algorithm is improved.
- We provide numerical results to verify the performance of the proposed algorithm. We show that, compared to the non-effective prediction (NEP) scheme, the proposed scheme with effective probability prediction (EP) consumes much less time for collecting the same percent of effective data blocks. Besides, the proposed algorithm adopting EP based on DNN shows performance gain against the baseline algorithm adopting EP based on Random Forests (RF).

The remainder of this paper is organized as follows: In Section 2, we give the system model of WS-WSN and in Section 3 we formulate the effective data block sum maximization problem. In Section 4, we introduce effective probability prediction and reformulate the problem. In Section 5, we present the joint UAV trajectories design, SN serve scheduling and UAV-SN association algorithm. In Section 6, we illustrate numerical results and validate the performance of the proposed algorithm. Finally, In Section 7, we conclude the paper.

2. System Model

2.1. UAV-Assisted Wide and Sparse WSN

As shown in Figure 1, we consider a UAV-assisted wide and sparse distributed wireless sensor network (WS-WSN). $M \geq 1$ rotary-wing UAVs are employed to collect data from $K \geq 1$ SNs distributed on a ground area. Assume the ground area is wide enough such that SNs are sparsely distributed, i.e., the distance between two SNs is large enough that they can not be simultaneously covered by one UAV. Denote the set of UAVs and the set of SNs by $\mathcal{M}$ with $|\mathcal{M}| = M$ and $\mathcal{K}$ with $|\mathcal{K}| = K$, respectively. The UAVs are assumed to fly at a fixed altitude H above the ground. Under a three-dimensional Cartesian coordinate system, the time-varying coordinate of UAV $m \in \mathcal{M}$ is denoted by $\mathbf{q}_m(t) = [x_m(t), y_m(t), H]^T \in \mathbb{R}^{3 \times 1}$. The exact location of SN $k \in \mathcal{K}$ is denoted by $\mathbf{s}_k = [x_k, y_k, 0]^T \in \mathbb{R}^{3 \times 1}$, which is assumed to be fixed.

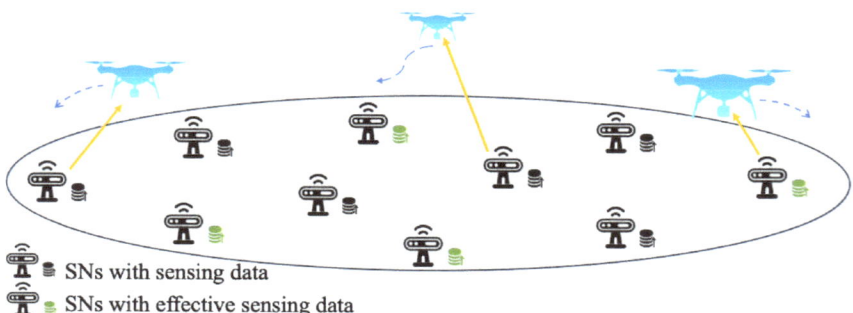

Figure 1. Indeterministic data collection in a UAV-assisted WS-WSN.

Assume that SNs sense the surrounding environment and produce sensing data periodically. Accordingly, the UAVs collect sensing data from SNs in a cycle mode, with the length of each flying period denoted as T. At the end of each period, UAVs return to depots for recharging and maintenance. Denote the locations of depots by $\mathbf{q}_m(0)$ $(m \in \mathcal{M})$. For ease of exposition, the period T is discretized into N equal time slots, with length $T_s = \frac{T}{N}$ chosen to be sufficiently small such that UAV locations are considered as approximately unchanged within each time slot even at the maximum flying speed $V_{\max}$. As a result, the flying trajectory of UAV m can be approximated by an N-length sequence

$\mathbf{q}_m[n] = [x_m[n], y_m[n], 0]^T$, $n = 1, \cdots, N$. During the n-th time slot, the distance between UAV m and SN k is $d_{m,k}[n] = \sqrt{\|\mathbf{q}_m[n] - \mathbf{s}_k\|^2 + H^2}$.

2.2. Indeterministic Data Collection

During each period, each SN senses its surrounding environment and produces a data block of D_0 bits. In this paper, considering the monitoring applications, the sensing data are generally referred to multimedia data such as images and videos. Hence, D_0 usually takes a value which is much larger than the size of traditional simple sensing data such as temperatures and humidness. Only a part of the sensing data blocks which contain target information are effective. For example, in wildlife monitoring applications, only the data blocks containing information related with wildlife activities are effective. For the l-th sensing period, use a binary variable $e_k^{(l)}$ to denote the effectiveness of sensing data block at SN k. For low cost and low energy consumption purpose, assume that SNs have no computing ability and are incapable to judge the effectiveness of sensing data. It means that the data effectiveness is indeterministic until it is collected and processed by a UAV. Besides, it is reasonable to assume that the number of UAVs is far smaller than the number of SNs, i.e., $M \ll K$. Thus, UAVs do not have enough time and energy to fly over and serve all of the SNs ergodically within a UAV flying period T.

2.3. Average Data Collection Rate

For the l-th flying period, define a set of binary variables $\{\alpha_{m,k}^{(l)}[n]\}$ to represent the association relationship between UAVs and SNs. $\alpha_{m,k}^{(l)}[n] = 1$ indicates that SN k is served by UAV m in the n-th time slot of the l-th flying period, otherwise $\alpha_{m,k}[n] = 0$. Under the wide and sparse WSN assumption, to facilitate the cooperation among multiple UAVs in order to cover more SNs, we assume that within a time slot, each SN is only served by at most one UAV, and a UAV serves no more than one SN. By this assumption, a simple communication protocol can satisfy the communication needs. There are many classic protocols that can be used, such as UAVCAN, IEEE 802.11, etc. The specific protocol design and implementation are not within the scope of this paper. Thus, we have two association constraints expressed as

$$\sum_{m=1}^{M} \alpha_{m,k}^{(l)}[n] \leq 1, \quad \forall l, n, k, \tag{1}$$

$$\sum_{k=1}^{K} \alpha_{m,k}^{(l)}[n] \leq 1, \quad \forall l, n, m. \tag{2}$$

Besides, considering that not all SNs are served by the UAV network, define a set of binary variables $\{\beta_k\}$ to represent whether an SN is served by UAVs during flying period l. Obviously, the association variables $\{\alpha_{m,k}^{(l)}[n]\}$ and the serve scheduling variables $\{\beta_k\}$ should satisfy

$$\sum_{n=1}^{N} \sum_{m=1}^{M} \alpha_{m,k}^{(l)}[n] \geq \beta_k^{(l)}, \quad \forall k. \tag{3}$$

We assume that the air-to-ground channels between UAVs and SNs are dominated by line-of-sight (LoS) channels [22]; in this paper, the NLoS components have limited impact on the transmission between the UAVs and SNs, and can be ignored for two reasons. First, in this paper, we consider the case that SNs being deployed in a wide and open area, such as farmland, or animal ecotope. With the minimum flying height limitation for UAVs, the probability of NLoS is relatively small. Second, the UAVs tend to fly to the locations above each SN in turn for better communication performance, making the probability of LoS larger. During time slot n, the channel power gain from UAV m to SN k is given by

$h_{m,k}[n] = \rho_0 d_{m,k}^{-2}[n] = \frac{\rho_0}{\|\mathbf{q}_m[n] - \mathbf{s}_k\|^2 + H^2}$, where ρ_0 is the channel power gain at a reference distance of 1 m (m). The received signal-to-interference-plus-noise ratio (SINR) at UAV m can be expressed as

$$\gamma_{m,k}[n] = \frac{P_s h_{m,k}[n]}{\sum\limits_{k'=1, k' \neq k}^{K} P_s h_{m,k'}[n] + \sigma^2}, \quad (4)$$

where P_s and σ^2 denote the transmit power of SNs and the Gaussian noise term, respectively. We assume that the UAVs collect data from SNs through the same time-frequency channel. Thus, a UAV receiver may experience interference from other SN-UAV transmission, with the interference power denoted as $\sum\limits_{k'=1, k' \neq k}^{K} P_s h_{m,k'}[n]$. The data transmission rate from SN k to UAV m in the time slot n in bits/second/Hertz (bps/Hz) is

$$R_{m,k}[n] = \log_2(1 + \gamma_{m,k}[n]). \quad (5)$$

The average data collection rate at SN k over the N time slots of a flying period is given by

$$R_k = \frac{1}{N} \sum_{n=1}^{N} \sum_{m=1}^{M} \alpha_{m,k}^{(l)}[n] R_{m,k}[n].$$

3. Effective Data Block Sum Maximization

For a wide and sparse WSN, UAVs are unable to serve all of the SNs within a flying period due to the energy limitation. We expect the UAV to collect as many effective data blocks as possible. Hence, we formulate a problem that maximizes the sum of effective sensing data blocks via jointly optimizing the UAV trajectories, SN serve scheduling as well as UAV-SN associations, formulated by

$$(\text{P1}): \max_{\mathbf{Q}, \mathbf{A}, \mathbf{B}} \sum_{k=1}^{K} e_k^{(l)} \beta_k^{(l)} \quad (6a)$$

s.t. (1), (2), (3),

$$\alpha_{m,k}^{(l)}[n] \in \{0, 1\}, \quad \forall m, k, n, \quad (6b)$$

$$\beta_k^{(l)} \in \{0, 1\}, \quad \forall k, \quad (6c)$$

$$N R_k \geq D_0, \quad \forall k \in \{\mathcal{K} \mid \beta_k^{(l)} = 1\}, \quad (6d)$$

$$\|\mathbf{q}_m[n] - \mathbf{q}_m[n-1]\| \leq \min(T_s V_{\max}, \Delta_{\max}), \quad \forall n, \forall m, \quad (6e)$$

$$\|\mathbf{q}_m[n] - \mathbf{q}_j[n]\| \geq d_{\text{safe}}, \quad \forall n, \forall m \neq j. \quad (6f)$$

where $\mathbf{Q} = \{\mathbf{q}_m[n], \forall m, n\}$, $\mathbf{A} = \{\alpha_{m,k}^{(l)}[n], \forall m, k, n\}$ and $\mathbf{B} = \{\beta_k^{(l)}, \forall k\}$ are variables of UAV trajectories, UAV-SN associations, and SN serve scheduling, respectively. The object function $\sum_{k=1}^{K} e_k^{(l)} \beta_k^{(l)}$ denotes the total number of effective sensing data blocks collected by UAVs within the l-th flying period. Constraint (6d) guarantees that for each SN served by UAVs, i.e., SN $k \in \{\mathcal{K} \mid \beta_k = 1\}$, the data block of D_0 bits can be completely collected by UAVs within a flying period. Constraint (6e) restricts both the UAV speed and the finite-sum approximation error introduced by the time discretization for UAV trajectories, with $V_{\max}$ and $\Delta_{\max}$ denoting the maximum UAV speed and the maximum discretization segment length, respectively. Constraint (6f) ensures collision avoidance between different UAVs with d_{safe} denoting the minimum inter-UAV distance. Problem (P1) is difficult to tackle since the data effectiveness indicators $e_k^{(l)}$ ($k \in \mathcal{K}$) contained in the objective function are indeterministic.

4. Effective Probability Prediction and Problem Reformulation

To make problem (P1) tractable, we first tackle the indeterministic indicators $e_k^{(l)}$. We predict the data's effective probability by a deep learning-based model, and reformulate (P1) into an effective probability-weighted data block sum maximization problem, the parameters of which are determinate.

4.1. Data Effective Probability Prediction

Define the effective probability of sensing data at SN k during the l-th flying period as

$$p_k^{(l)} = P\left(e_k^{(l)} = 1\right). \tag{7}$$

Use $\mathbf{P}^{(l)} = \{p_k^{(l)}\}_{k=1}^K$ to denote the effective probability vector of all SNs. Considering that in essence, $p_k^{(l)}$ is the probability that an event occurs. For example, in fire monitoring, $p_k^{(l)} = 1$ is the probability of a fire happening near SN k during period l. In wildlife monitoring, $p_k^{(l)} = 1$ is the probability of animals passing by SN k during period l. Hence, the values of $\mathbf{P}^{(l)}$ are highly relative to the features related to the event and can be predicted by relevant feature data.

We construct an effective probability prediction model, aiming to approximate a function mapping $\varphi : X \to Y$, where X is a set of input variables including the feature data of SNs such as position coordinates, temperature, humidity, etc. Y is the output variable denoting the effective probability of sensing data at SNs. It should be specifically noted that the environmental perception data mentioned above are relatively small in volume, typically several tens to several hundred bytes, and can be quickly and easily obtained within a few minutes before the formal sensing data collection process begins. For example, a UAV can be dispatched to fly along a circular trajectory that covers the area to collect the environmental perception data. For a 0.5×0.5 km^2 square area, it only takes 40 s to fly around the area when the UAV is flying at a speed of 50 m/s. After collecting the environmental perception data, the UAV returns to the control center. Then, the control center would predict the effectiveness of the sensing data and plan the UAVs' trajectories for sensing data collection. Compared to the environmental perception data, the sensing data used for detection and recognition tasks, such as surveillance video data, are relatively larger in volume, reaching gigabytes (Gb). UAVs are required to fly close to the SNs and hover for a long period to collect sensing data. It is inefficient and excessively energy-consuming for UAVs to collect sensing data from all of the SNs. Therefore, we need to predict the effectiveness of sensing data based on the environmental perception data and optimize the collection efficiency.

A deep learning-based modeling framework is adopted. The training dataset is

$$T_I = \{(x_{k,i}, y_{k,i}) : k \in \mathcal{K}, i \in [1, I])\}, \tag{8}$$

where I is the number of records in the training dataset, $x_{k,i}$ and $y_{k,i}$ are, respectively, the training inputs and target output of the i-th record related to SN k. The training determines the parameters of the model that minimize the loss function between the estimated and real values of the output variables, yielding a non-linear interpolation-based input-output mapping φ. Hence, the matrix $\mathbf{P}^{(l)}$ can be predicted by

$$\hat{\mathbf{P}}^{(l)} = \varphi(X^{(l)}). \tag{9}$$

4.2. Problem Reformulation

By substituting the effectiveness indicators $e_k^{(l)}$ with the predicted effective probability $p_k^{(l)}$, problem (P1) can be reformulated into an expected effective data block sum maximization problem (EE-SMP), which aims to maximize the expected sum of effective data blocks, given by

$$(\text{P2}) : \max_{\mathbf{Q},\mathbf{A},\mathbf{B}} \sum_{k=1}^{K} p_k \beta_k \tag{10}$$

s.t. (1), (2), (3), (6b), (6c), (6d), (6e), (6f).

The subscript l is omitted.

Solving the reformulated problem (P2) is still challenging, since (P2) is quite different from traditional joint UAV trajectories design and UAV-SN association optimization problems. In particular, the number of constraints in (P1) given by (6d) varies with its optimization variable, i.e., the UAV-SN scheduling indicator $\mathbf{B} = \{\beta_k^{(l)}, \forall k\}$, since (6d) only restricts the SNs that are served by UAVs.

5. Modified Knapsack Algorithm

The EE-SMP given by (P2) is a complicate combinational optimization problem. Besides, (P2) contains varying constraints, i.e., the constraint given by (6d) varies with the specific value of the serve scheduling variables $\mathbf{B} = \{\beta_k, \forall k\}$. It means that, if an SN k is chosen to be served by UAV, i.e., $\beta_k = 1$, then the lower bound of its average transmission rate R_k should be limited to ensure the collection of a complete data block. Otherwise, for a non-served SN, there is no minimum data collection rate limitation. Hence, problem (P2) can not be treated as a traditional optimization problem.

We solve the EE-SMP by modeling it as a modified knapsack problem. First, we treat each sensor node, named SN k, as an item, with its data effective probability p_k being viewed as the value of the item. Then, we create a virtual knapsack to accommodate the items. Maximizing the expected effective data block sum $\sum_{k=1}^{K} p_k \beta_k$ via optimizing the serve scheduling strategy $\mathbf{B} = \{\beta_k, \forall k\}$, as given in (P2), is equivalent to finding a packing strategy $\mathbf{B}$ that maximizes the total value of the virtual knapsack. Hence, the EE-SMP can be equivalently reformulated into a modified knapsack problem, given by

$$(\text{P3}) : \max_{\tilde{\mathbf{B}}} \sum_{k=1}^{K} p_k \tilde{\beta}_k \tag{11a}$$

$$\text{s.t.} \quad \tilde{\beta}_k \in \{0,1\}, \quad \forall k, \tag{11b}$$

$$(\text{P2})|_{\mathbf{B}=\tilde{\mathbf{B}}} \text{ is feasible.} \tag{11c}$$

(11c) means that the packing strategy $\tilde{\mathbf{B}} = \{\tilde{\beta}_k, \forall k\}$ should be feasible to (P2) in order to make it solvable. (P3) is a modified knapsack problem (MKP) with the capacity constraint in traditional knapsack problem being replaced with a feasibility constraint given by (11c). To solve (P3), we first solve the feasibility problem of (P2) with given $\mathbf{B}$, i.e., $(\text{P2})|_{\mathbf{B}=\tilde{\mathbf{B}}}$, and then we resort to the greedy algorithm to solve the MKP.

5.1. the Feasibility Problem of $(\text{P2})|_{\mathbf{B}=\tilde{\mathbf{B}}}$

Given $\mathbf{B} = \tilde{\mathbf{B}}$, the feasibility problem of (P2) can be expressed as

$$(\text{P4}) : \text{find} \quad \mathbf{Q}, \mathbf{A} \tag{12a}$$

$$\text{s.t.} \quad \sum_{n=1}^{N} \sum_{m=1}^{M} \alpha_{m,k}[n] \geq \tilde{\beta}_k, \quad \forall k, \tag{12b}$$

$$NR_k \geq D_0, \quad \forall k \in \{\mathcal{K} \mid \tilde{\beta}_k = 1\}, \tag{12c}$$

$$(1), (2), (6e), (6f). \tag{12d}$$

(P4) can be further transformed into a max-min rate problem given by

$$(\text{P5}) : \max_{\tilde{\mathbf{A}}, \mathbf{Q}, \eta} \eta \tag{13a}$$

$$\text{s.t.} \quad NR_k \geq \eta, \quad \forall k \in \tilde{\mathcal{K}}, \tag{13b}$$

$$(1), (2), (6e), (6f). \tag{13c}$$

where $\tilde{\mathbf{A}} = \{\mathbf{A} | \sum_{n=1}^{N} \sum_{m=1}^{M} \alpha_{m,k}[n] \geq \tilde{\beta}_k\}$ and $\tilde{\mathcal{K}} = \{\mathcal{K} \mid \tilde{\beta}_k = 1\}$ are the valid UAV-SN association variables and the set of served SNs given $\mathbf{B} = \tilde{\mathbf{B}}$, respectively. (P4) is feasible if and only if the optimal object value of (P5), denoted by η^{opt}, satisfying $\eta^{\text{opt}} \geq D_0$. (P5) is a classic joint UAV trajectory and UAV-SN associations optimization problem. We decompose it into two sub-problems, and resort to the block coordinate decent (BCD) and successive convex approximation (SCA) techniques. The two sub-problems are alternately solved until the algorithm converges, as referred in [29].

5.2. Algorithm Design

Based on the solution to the feasibility problem of (P2), a hybrid greedy algorithm is proposed to resolve the modified knapsack algorithm, as summarized in Algorithm 1.

Algorithm 1 Hybrid greedy algorithm for (P3)

1: **repeat**
2:    Initialize iteration index $r = 0$, set item buffer $\mathcal{I}^0 = \mathcal{K}$, set knapsack pack $\mathbf{B}^0 = 0$ and knapsack value $v^0 = 0$.
3:    For the i-th iteration, find the item k^r which has the maximum value in $\mathcal{I}^r$. Take k^r out from $\mathcal{I}^r$, and set $\mathcal{I}^{r+1} = \mathcal{I}^r - k^r$.
4:    Set $\tilde{\mathbf{B}} = \mathbf{B}^r$.
5:    Update $\tilde{\mathbf{B}}$ by setting $\tilde{\beta}_{k^r} = 1$.
6:    Check the feasibility of $(\text{P2})|_{\mathbf{B}=\tilde{\mathbf{B}}}$ via solving (P5).
7:    **if** $(\text{P2})|_{\mathbf{B}=\tilde{\mathbf{B}}}$ is feasible **then**
8:       Put item k^r into knapsack with updating $\mathbf{B}^{r+1} = \tilde{\mathbf{B}}$.
9:    **else**
10:      Discard item k^r with updating $\mathbf{B}^{r+1} = \mathbf{B}^r$.
11:      Go back to step 3.
12:    **end if**
13:    update $r = r + 1$.
14: **until** $\mathcal{I}^{r+1}$ is empty.
15: Output $\mathbf{B}^{r+1}$ as the optimal knapsack packing strategy.

---

6. Numerical Results

In this section, we provide numerical results to demonstrate the effectiveness of the proposed scheme. We consider a scenario where the sensing nodes are sparsely distributed, where $K = 20$ SNs are randomly and uniformly distributed within a 2D square area of 0.5×0.5 km^2. Two UAVs are employed to collect data from ground SNs. For the effective probability prediction model training, a natural area sensing dataset from a Chinese telecom operator is adopted. As part of research collaborative efforts, we were granted access to the labeled data collected by 20 sensors monitoring a forestry area in 2022. The total number of sensing records is 24,612. For each sensing record, the input variables include position coordinates, altitudes, temperature, humidness, wind strength, pressure, light, and sound volume, and the output is a Boolean variable indicating the data effectiveness. A two-layer DNN network is used to train the prediction model;compared to the traditional machine learning algorithms like decision trees [30] and linear models [31], DNN is distinguished by deep, hierarchical architecture, which facilitates the learning of complex, nonlinear patterns within data. When dealing with predictive tasks, DNNs often demonstrate superior predictive performance. The main simulation setups are summarized in Table 1.

Table 1. Simulation parameters.

Parameter	Description	Value
T	Flying period of UAV	20–100 s
D_0	Sensing data block size	2–14 Mbits
ρ_0	Channel power gain at the reference distance 1 m	−30 dB
σ^2	Receiver noise power	−110 dBm
$\Delta_{\max}$	Maximum discretization segment length	5 m
$V_{\max}$	Maximum UAV speed	20 m/s
D_{safe}	Minimum inter-UAV distance to ensure collision avoidance	50 m
T_s	Length of discretized time slot	1 s
P_s	Transmit power of SN	5 mW
H	Flying altitude of UAVs	80 m

Two baseline schemes are adopted for comparison. (1) MKP with effective probability predicted by Random Forests (EP-RF). (2) UAVs fly along a Traveling Salesman route without effective probability prediction (NEP). In particular, UAVs determine whether to fly to the nearest SN to collect data, or fly back to depots to recharge, according to remaining battery energy.

6.1. DNN-Based Effective Probability Prediction

The variables of each record for model training include position coordinates (X_1, X_2), altitudes (X_3), temperature (X_4), humidness (X_5), wind strength (X_6), pressure (X_7), light (X_8), "sound volume" (X_9) and a Boolean variable indicating whether each SN obtains an effective sensing data block (Y). We normalize all continuous variables by Z-Score normalization, formulated as

$$\tilde{x}_i^{(j)} = (x_i^{(j)} - \mu_i)/\sigma_i, \quad \forall i, j. \tag{14}$$

where $x_i^{(j)}$ is the value of variable X_i of the j-th record, μ_i and σ_i are the mean and the standard deviation of X_i, respectively. The dataset has been used for training and testing in a 90:10 ratio.

A DNN network has been used to train the prediction model, which contains an input layer, two hidden layers and an output layer, as shown in Figure 2. We use normalized data $X_1, X_2, \cdots, X_9$ as the inputs of the DNN network and Y as the output. Through repeated testing and verification, it was found that the two hidden layers with 50 neurons in each layer were able to provide enough nonlinear expression ability. We use the ReLU function to process the input of the hidden layer neurons, and the Sigmoid function to process the output.

We conduct training and performance tests on Intel Core i7-8569 and 16G RAM. The number of epochs and batch size are set as 500 and 128, respectively. The Adam optimizer with a learning rate of 0.01 is used. Cross entropy is adopted as the loss function, and 10% of the training data have been used as a validation set, which has been used for model selection and hyperparameter optimization. The validation set has provided an unbiased evaluation of a model fit during the training phase as it was not used in the training process itself. The loss value of training and validation at successive epochs is shown in Figure 3. When the model training has been completed, the test set is used to evaluate the final performance of the model, which has been completely unseen by the model during both the training and validation phases. We evaluated the prediction model by ROC curve, and performance evaluation on the test set is shown in Figure 4. The area under the curve (AUC) is 0.953.

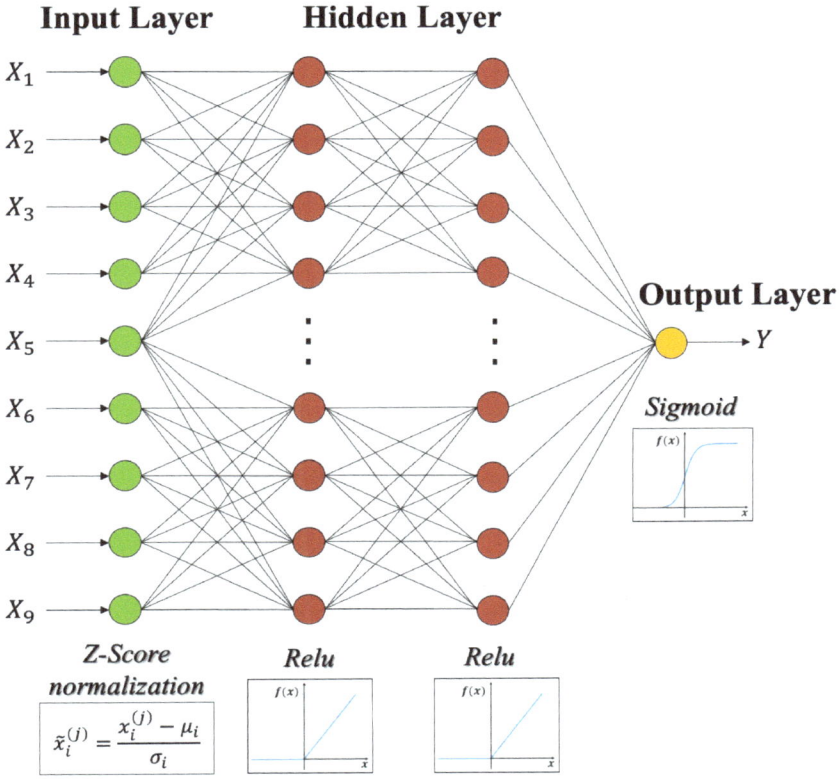

Figure 2. The DNN network of the data effective probability prediction model.

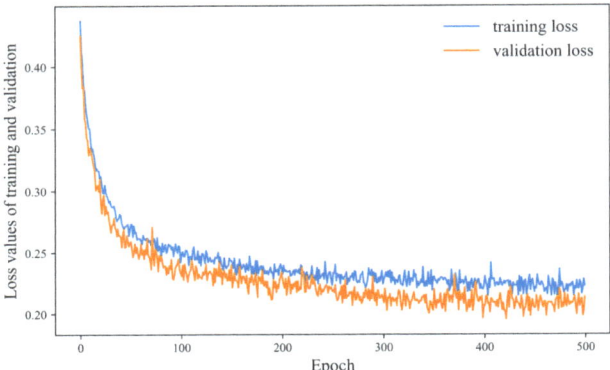

Figure 3. The loss value of training and validation at successive epochs.

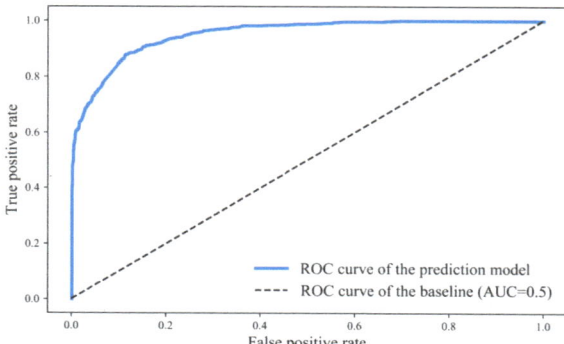

Figure 4. The ROC curve of performance evaluation on test dataset.

6.2. Illustrations of Optimized UAV Trajectories and SN Serve Scheduling

Figures 5 and 6 give typical results of optimized UAV trajectories and SN serve scheduling with the three schemes in one implementation, when the UAV flying period T set to be 40 s and 80 s, respectively. It is observed that, with the two schemes with effective probability prediction, i.e., EP-DNN and EP-RF, UAVs tend to choose SNs of greater predicted effective probability, which may be located far away from the depots. Moreover, among the two EP-based schemes, UAVs with the EP-DNN scheme show much more accuracy via choosing more effective SNs, since DNN performs better than RF in prediction accuracy. In contrast, with the NEP scheme, UAVs give preference to the nearby SNs around the depots, and collect more ineffective data than the other two schemes. Comparing Figures 5 and 6, it shows that with a larger flying period T, the EP-based schemes almost cover all of the effective SNs. While with the NEP scheme, a high proportion of sensing data are collected from ineffective SNs.

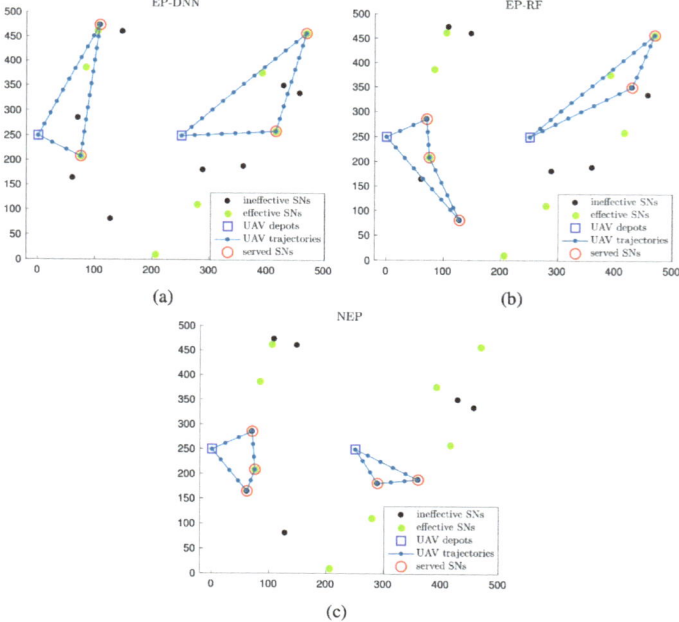

Figure 5. The optimized UAV trajectories and SN serve scheduling of the EP-DNN scheme (**a**), the EP-RF scheme (**b**) and the NEP scheme (**c**) when $T = 40$ s.

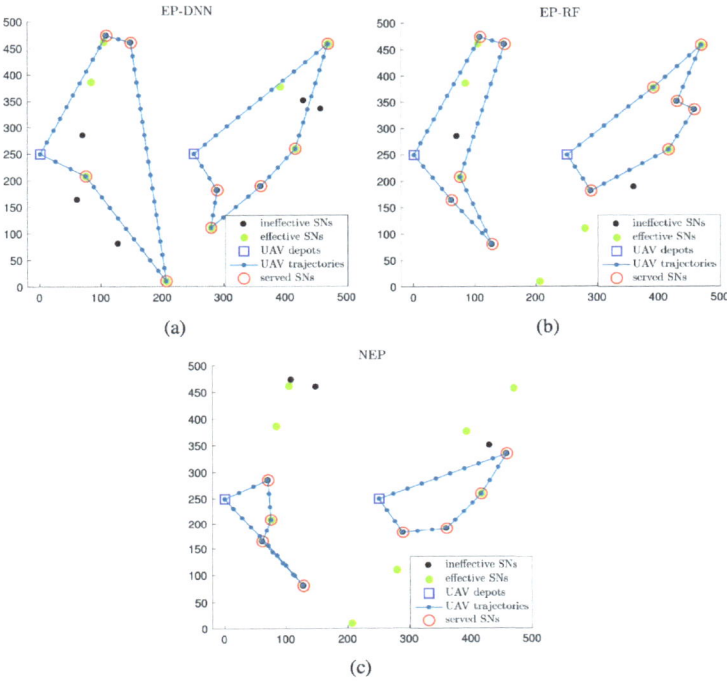

Figure 6. The optimized UAV trajectories and SN serve scheduling of the EP-DNN scheme (**a**), the EP-RF scheme (**b**) and the NEP scheme (**c**) when $T = 80$ s.

6.3. Collected Data Block Ratios

To compare the performance of different schemes, we consider two metrics regarding data collection performance. One is the collection ratio of all data blocks (DR), i.e., the ratio of the number of collected data blocks and the total number of data blocks. The other is the collection ratio of effective data blocks (EDR), i.e., the ratio of the number of collected effective data blocks and the total number of effective data blocks. Figure 7 shows the curves of DR (plotted by blue and dotted curves) and EDR (plotted by red and solid curves) with respect to the flying period T. When T is small, the EP-DNN, EP-RF and NEP schemes show similar performance in both DR and EDR, since the period time is only enough for UAVs flying around depots. As T increases, the two EP-based schemes, exhibit advantages in both DR and EDR against the NEP scheme, with the performance gaps also increasing. In particular, the DR performance of EP-DNN and EP-RF is similar, with the gain compared to NEP coming from joint trajectories and serve scheduling optimization. While the EDR performance of EP-DNN is better than EP-RF, due to the prediction accuracy gain of EP-DNN against EP-RF. When T is large enough, e.g., taking values of 90 s or 100 s, the performance curves of EP-DNN and EP-RF converges, since in this case, UAVs are able to serve almost all of the SNs. For a quantitative comparison, consider 50% effective data collection, the NEP scheme takes near 80 s, while the EP-DNN and EP-RF schemes take 52 s and 60 s, corresponding to near 35% and 25% time reduction, respectively.

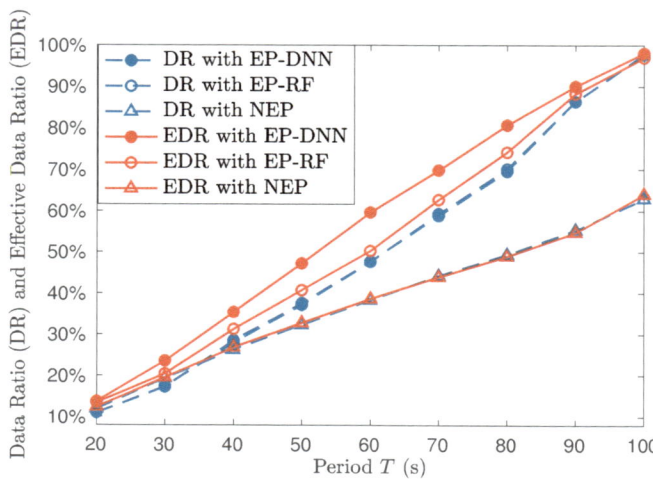

Figure 7. The collection ratio of data blocks (DR) and effective data blocks (EDR) with respect to the flying period T.

Figure 8 shows the curves of DR (plotted by blue and dotted curves) and EDR (plotted by red and solid curves) with respect to the data block size D_0 given $T = 80$ s. When D_0 is small, UAVs consume less time in data collection from SNs and remain more time for UAVs to fly and visit more SNs. In this case, the EP-DNN scheme and the EP-RF scheme show obvious performance gains in both DR and EDR compared to the NEP scheme. With D_0 increases, it is observed that the advantages of the two EP schemes reduce, since the time remaining for UAVs to fly decreases, and hence the space for trajectory optimization decreases. Besides, it is worth noticing that, as D_0 increases, the gap between EP-DNN and EP-RF becomes first large and then stable. The reason is that, when D_0 is small, UAVs are able to visit more SNs, and probabilistically include more effective SNs, hence the prediction accuracy becomes less important.

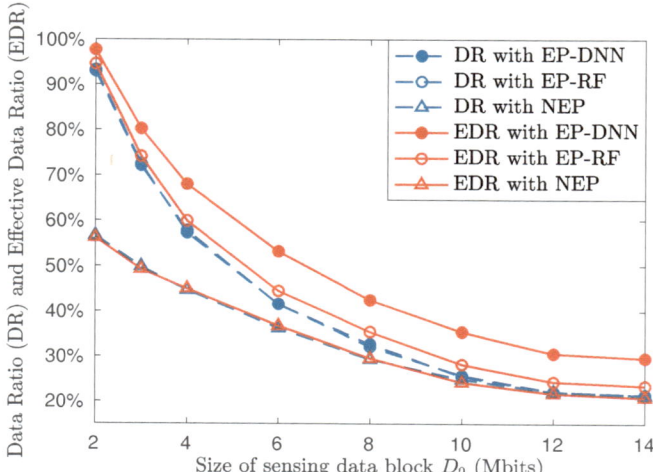

Figure 8. DR and EDR with respect to the data block size D_0, with T set as 80 s.

7. Conclusions

In this paper, we focus on an indeterministic data collection task for UAV-assisted wide and sparse wireless sensor networks, where the SNs obtain effective data randomly, and the UAV has no pre-knowledge about which sensor has effective data. We jointly optimized the UAV trajectories, SN serve scheduling and UAV-SN associations to maximize the sum of collected effective sensing data blocks. We addressed the indeterministic effective indicator by introducing an effectiveness probability prediction model and tackled the issue caused by a varying number of constraints by proposing a two-layer modified knapsack algorithm, within which a feasibility problem is resolved iteratively to find the optimal packing strategy. Numerical results demonstrate that the proposed scheme had remarkable advantages in the sum of effective data blocks, and reduced the completion time for collecting the same ratio of effective data by nearly 30%.

Author Contributions: Conceptualization, Y.D. and Y.G.; methodology, Y.D.; software, J.H.; validation, J.H. and Z.C.; formal analysis, Y.D. and Z.C.; investigation, Y.D. and Z.C.; resources, Y.D.; data curation, Z.C.; writing—original draft preparation, Y.D.; writing—review and editing, Y.G.; visualization, Y.D.; supervision, Y.G.; project administration, Y.G. All authors have read and agreed to the published version of the manuscript.

Funding: This research received no external funding.

Institutional Review Board Statement: Not applicable.

Informed Consent Statement: Not applicable.

Data Availability Statement: Data are contained within the article.

Conflicts of Interest: The authors declare no conflict of interest.

References

1. Shahid, M.; Tariq, M.; Iqbal, Z.; Albarakati, H.M.; Fatima, N.; Khan, M.A.; Shabaz, M. Link-quality based energy-efficient routing protocol for WSN in IoT. *IEEE Trans. Consum. Electron.* **2024**, *70*, 4645–4653. [CrossRef]
2. Shen, X.; Yi, B.; Liu, H.; Zhang, W.; Zhang, Z.; Liu, S.; Xiong, N. Deep variational matrix factorization with knowledge embedding for recommendation system. *IEEE Trans. Knowl. Data Eng.* **2019**, *33*, 1906–1918. [CrossRef]
3. Lin, C.; Xiong, N.; Park, J.H.; Kim, T.H. Dynamic power management in new architecture of wireless sensor networks. *Int. J. Commun. Syst.* **2009**, *22*, 671–693. [CrossRef]
4. Lin, C.; He, Y.X.; Xiong, N. An energy-efficient dynamic power management in wireless sensor networks. In Proceedings of the 2006 Fifth International Symposium on Parallel and distributed computing, Timisoara, Romania, 6–9 July 2006; pp. 148–154.
5. Guo, W.; Xiong, N.; Vasilakos, A.V.; Chen, G.; Yu, C. Distributed k–connected fault–tolerant topology control algorithms with PSO in future autonomic sensor systems. *Int. J. Sens. Netw.* **2012**, *12*, 53–62. [CrossRef]
6. Shu, L.; Zhang, Y.; Yu, Z.; Yang, L.T.; Hauswirth, M.; Xiong, N. Context-aware cross-layer optimized video streaming in wireless multimedia sensor networks. *J. Supercomput.* **2010**, *54*, 94–121. [CrossRef]
7. Yang, Y.; Xiong, N.; Chong, N.Y.; Defago, X. A Decentralized and Adaptive Flocking Algorithm for Autonomous Mobile Robots. In Proceedings of the 2008 The 3rd International Conference on Grid and Pervasive Computing-Workshops, Kunming, China, 25–28 May 2008; pp. 262–268.
8. Zhang, W.; Zhu, S.; Tang, J.; Xiong, N. A novel trust management scheme based on Dempster–Shafer evidence theory for malicious nodes detection in wireless sensor networks. *J. Supercomput.* **2018**, *74*, 1779–1801. [CrossRef]
9. Sang, Y.; Shen, H.; Tan, Y.; Xiong, N. Efficient protocols for privacy preserving matching against distributed datasets. In Proceedings of the Information and Communications Security: 8th International Conference, Raleigh, NC, USA, 4–7 December 2006; pp. 210–227.
10. Xu, W.; Fang, W.; Ding, Y.; Zou, M.; Xiong, N. Accelerating federated learning for iot in big data analytics with pruning, quantization and selective updating. *IEEE Access* **2021**, *9*, 38457–38466. [CrossRef]
11. Shen, Y.; Fang, Z.; Gao, Y.; Xiong, N.; Zhong, C.; Tang, X. Coronary arteries segmentation based on 3D FCN with attention gate and level set function. *IEEE Access* **2021**, *7*, 42826–42835. [CrossRef]
12. Wei, Z.; Zhu, M.; Zhang, N.; Wang, L.; Zou, Y.; Meng, Z.; Feng, Z. UAV-assisted data collection for Internet of Things: A survey. *IEEE Internet Things J.* **2022**, *9*, 15460–15483. [CrossRef]
13. Zhang, L. Joint Energy Replenishment and Data Collection Based on Deep Reinforcement Learning for Wireless Rechargeable Sensor Networks. *IEEE Trans. Consum. Electron.* **2023**, *70*, 1052–1062. [CrossRef]
14. Liu, X.; Song, H.; Liu, A. Intelligent UAVs trajectory optimization from space-time for data collection in social networks. *IEEE Trans. Netw. Sci. Eng.* **2020**, *8*, 853–864. [CrossRef]

15. Li, P.; Xu, J. Fundamental rate limits of UAV-enabled multiple access channel with trajectory optimization. *IEEE Trans. Wirel. Commun.* **2020**, *19*, 458–474. [CrossRef]
16. Wu, Q.; Zhang, R. Common throughput maximization in UAV-enabled OFDMA systems with delay consideration. *IEEE Trans. Commun.* **2018**, *66*, 6614–6627. [CrossRef]
17. Chen, W.; Zhao, S.; Zhang, R.; Chen, Y.; Yang, L. UAV-assisted data collection with nonorthogonal multiple access. *IEEE Internet Things J.* **2021**, *8*, 501–511. [CrossRef]
18. Gu, J.; Wang, H.; Ding, G.; Xu, Y.; Xue, Z.; Zhou, H. Energy-constrained completion time minimization in UAV-enabled Internet of Things. *IEEE Internet Things J.* **2020**, *7*, 5491–5503. [CrossRef]
19. Wang, W.; Zhao, N.; Chen, L.; Liu, X.; Chen, Y.; Niyato, D. UAV-assisted time-efficient data collection via uplink NOMA. *IEEE Trans. Commun.* **2021**, *69*, 7851–7863. [CrossRef]
20. Yuan, X.; Hu, Y.; Zhang, J.; Schmeink, A. Joint user scheduling and UAV trajectory design on completion time minimization for UAV-aided data collection. *IEEE Trans. Wirel. Commun.* **2023**, *22*, 3884–3898. [CrossRef]
21. Li, M.; Liu, X.; Wang, H. Completion time minimization considering GNs' energy for UAV-assisted data collection. *IEEE Wirel. Commun. Lett.* **2023**, *12*, 2128–2132. [CrossRef]
22. Sun, C.; Xiong, X.; Ni, W.; Wang, X. Three-Dimensional Trajectory Design for Energy-Efficient UAV-Assisted Data Collection. In Proceedings of the 2022 IEEE International Conference on Communications, Seoul, Republic of Korea, 16–20 May 2022; pp. 3580–3585.
23. Li, P.; Chai, R.; Tang, R.; Pu, R. Long Term Energy Consumption Minimization-based Data Collection for UAV-Assisted WSNs. In Proceedings of the 2023 IEEE 98th Vehicular Technology Conference, Hong Kong, China, 10–13 October 2023; pp. 1–5.
24. Chen, J.; Yan, F.; Mao, S.; Shen, F.; Xia, W.; Wu, Y.; Shen, L. Efficient Data Collection in Large-Scale UAV-aided Wireless Sensor Networks. In Proceedings of the 2019 11th International Conference on Wireless Communications and Signal Processing, Xi'an, China, 23–25 October 2019; pp. 1–5.
25. Chai, R.; Gao, Y.; Sun, R.; Zhao, L.; Chen, Q. Time-Oriented Joint Clustering and UAV Trajectory Planning in UAV-Assisted WSNs: Leveraging Parallel Transmission and Variable Velocity Scheme. *IEEE Trans. Intell. Transp. Syst.* **2023**, *24*, 12092–12106. [CrossRef]
26. Li, D.; Xu, S.; Zhao, C.; Wang, Y.; Xu, R.; Ai, B. Data Collection In Laser-Powered UAV-Assisted IoT Networks: Phased Scheme Design Based on Improved Clustering Algorithm. *IEEE Trans. Green Commun. Netw.* **2024**, *8*, 482–497. [CrossRef]
27. Liu, X.; Liu, H.; Zheng, K.; Liu, J.; Taleb, T.; Shiratori, N. AoI-minimal Clustering, Transmission and Trajectory Co-design for UAV-assisted WPCNs. *IEEE Trans. Veh. Technol.* **2024**. [CrossRef]
28. Zhu, B.; Bedeer, E.; Nguyen, H.H.; Barton, R.; Gao, Z. UAV Trajectory Planning for AoI-Minimal Data Collection in UAV-Aided IoT Networks by Transformer. *IEEE Trans. Wirel. Commun.* **2023**, *22*, 1343-1358. [CrossRef]
29. Zeng, Y.; Zhang, R. Energy-efficient UAV communication with trajectory optimization. *IEEE Trans. Wirel. Commun.* **2017**, *16*, 3747–3760. [CrossRef]
30. Moraru, A.; Pesko, M.; Porcius M. Using machine learning on sensor data. *J. Comput. Inf. Technol.* **2010**, *18*, 341-347. [CrossRef]
31. Soleymani S.A.; Goudarzi, S.; Kama, N. A hybrid prediction model for energy-efficient data collection in wireless sensor networks. *Symmetry* **2024**, *12*, 2024. [CrossRef]

Disclaimer/Publisher's Note: The statements, opinions and data contained in all publications are solely those of the individual author(s) and contributor(s) and not of MDPI and/or the editor(s). MDPI and/or the editor(s) disclaim responsibility for any injury to people or property resulting from any ideas, methods, instructions or products referred to in the content.

MDPI AG
Grosspeteranlage 5
4052 Basel
Switzerland
Tel.: +41 61 683 77 34

Sensors Editorial Office
E-mail: sensors@mdpi.com
www.mdpi.com/journal/sensors

Disclaimer/Publisher's Note: The title and front matter of this reprint are at the discretion of the Guest Editors. The publisher is not responsible for their content or any associated concerns. The statements, opinions and data contained in all individual articles are solely those of the individual Editors and contributors and not of MDPI. MDPI disclaims responsibility for any injury to people or property resulting from any ideas, methods, instructions or products referred to in the content.

www.ingramcontent.com/pod-product-compliance
Lightning Source LLC
LaVergne TN
LVHW072350090526
838202LV00019B/2516